T0201070

# An Introduction to
# Wavelet Modulated
# Inverters

A complete list of titles in the IEEE Press Series on Power Engineering appears at
the end of this book.

# An Introduction to Wavelet Modulated Inverters

S. A. Saleh
M. Azizur Rahman

IEEE
PRESS
SERIES
ON POWER
ENGINEERING

Mohamed E. El-Hawary, *Series Editor*

IEEE PRESS

A John Wiley & Sons, Inc., Publication

Published by John Wiley & Sons, Inc., Hoboken, New Jersey.
Published simultaneously in Canada.

For general information on our other products and services or for technical support, please contact our Customer Care Department within the United States at (800) 762-2974, outside the United States at (317) 572-3993 or fax (317) 572-4002.

Wiley also publishes its books in a variety of electronic formats. Some content that appears in print may not be available in electronic formats. For more information about Wiley products, visit our web site at www.wiley.com.

*Library of Congress Cataloging-in-Publication Data:*

An introduction to wavelet modulated inverters / S. A. Saleh and M. Azizur Rahman.
    p. cm.—(IEEE Press series on power engineering)
    ISBN 978-0-470-61048-0 (cloth)
  1. Electric inverters.  2. Wavelets (Mathematics)  3. Modulation (Electronics)
I S. A. Saleh,  II M. Azizur Rahman
  TK7872.I65R54 2010
  621.3815'322—dc22

                                           2010004248

Printed in the United States of America.

10 9 8 7 6 5 4 3 2 1

# Contents

**Preface**    ix

**List of Symbols**    xi

**List of Abbreviations**    xv

**1. Introduction to Power Inverters**    1

  1.1  Fundamental Inverter Topologies    1
      1.1.1  Single-Phase ($1\phi$) Inverters    2
      1.1.2  Three-Phase ($3\phi$) Inverters    4
  1.2  Multilevel Inverter Topologies    6
      1.2.1  Neutral-Point Clamped Multilevel Inverter    7
      1.2.2  Diode-Clamped Multilevel Inverter    8
      1.2.3  Capacitor-Clamped Multilevel Inverter    8
      1.2.4  Cascaded $H$-Bridge Multilevel Inverter    9
  1.3  Fundamental Inverter Switching    11
  1.4  Harmonic Distortion    15
  1.5  Summary    17

**2. Wavelets and the Sampling Theorem**    19

  2.1  Introduction    19
  2.2  Wavelet Basis Functions    21
      2.2.1  Orthogonal Wavelet Basis Functions    23
      2.2.2  Semi-Orthogonal Wavelet Basis Functions    25
      2.2.3  Bi-Orthogonal Wavelet Basis Functions    27
      2.2.4  Shift-Orthogonal Wavelet Basis Functions    28
  2.3  Sampling Process as a Multiresolution Analysis (MRA)    29
  2.4  Sampling Forms    33
      2.4.1  Uniform Sampling    33
      2.4.2  Nonuniform Sampling    35
      2.4.3  Nonuniform Recurrent Sampling    36
  2.5  Wavelet Sampling Theory    37
  2.6  Summary    39

**3. Modeling of Power Inverters**    **41**

  3.1   Introduction    41
  3.2   Sampling-Based Modeling of Single-Phase Inverters    43
        3.2.1   Nonuniform Sampling-Based Representation    44
        3.2.2   Reconstructing the Reference-Modulating
                Signal from Nonuniform Samples    46
  3.3   Testing the Nonuniform Recurrent Sampling-Based Model of
        Inverters    51
        3.3.1   PWM Inverter Output Voltage for Two Carrier
                Frequencies    52
  3.4   Sampling-Based Modeling of Three-Phase Inverters    53
  3.5   Summary    62

**4. Scale-Based Linearly Combined Wavelets**    **65**

  4.1   Introduction    65
  4.2   Scale-Based Linearly Combined Wavelet Basis Functions    66
        4.2.1   Balancing the Order of the Scale-Based Linearly Combined
                Scaling Function $\varphi(t)$    70
        4.2.2   Scale-Based Linearly Combined Wavelet Function $\psi_\varphi(t)$    72
        4.2.3   Construction of Scale-Based Linearly Combined Synthesis
                Scaling Functions $\tilde{\varphi}(t)$    74
  4.3   Nondyadic MRA Structure    76
        4.3.1   MRA for Nonuniform Recurrent Sampling    76
  4.4   Scale-Based Linearly Combined Scaling Functions for Three-Phase
        Inverters    79
  4.5   Summary    83

**5. Single-Phase Wavelet Modulated Inverters**    **85**

  5.1   Introduction    85
  5.2   Implementing the Wavelet Modulation Technique    85
  5.3   Simulated Performance of a Wavelet Modulated Inverter    88
  5.4   Experimental Performance of a Wavelet Modulated Inverter    95
  5.5   The Scale-Time Interval Factor $\gamma$    101
  5.6   Summary    106

**6. Three-Phase Wavelet Modulated Inverters**    **107**

  6.1   Introduction    107
  6.2   Implementing the Wavelet Modulation Technique for a Three-Phase
        Inverter    108
  6.3   Simulated Performance of a Three-Phase Wavelet Modulated
        Inverter    111
  6.4   Experimental Performance of a Three-Phase Wavelet Modulated
        Inverter    119
  6.5   Summary    127

**Appendix A    Nondyadic MRA for 3$\phi$ WM Inverters**                                    **131**

  A.1    Preliminary Derivations        131
  A.2    Time and Scale Localization of MRA Spaces        132

**Bibliography**        135

**Index**      143

# Preface

Power electronics has emerged as the enabling technology for industrial competitiveness in the world marketplace in the $21^{st}$ century. One of the most valuable achievements in power electronics is the introduction of degree of freedom to variable frequency in order to produce electric power with the least environmental consequences, keep the power system stable, improve products and services, and minimize energy wastage, as well as enhance power quality and improve efficiency.

The industry standard approach has been the carrier-based pulse width modulation (PWM), which is based on generating the switching pulses through comparing a high-frequency carrier signal with a low-frequency sinusoidal reference modulating signal. The general objective of dispensing with the carrier signal altogether has been successfully realized by the non-uniform sampling-reconstruction process, which is supported by a non-dyadic multiresolution analysis structure of the modulating signal. Such a dc-ac power electronics converter is called the wavelet modulated (WM) inverter. The salient feature of the WM inverter is its ability to increase the magnitude of the fundamental component and to reduce the harmonic components from their outputs. This feature is evident from close-to-ideal sampling reconstruction of the sinusoidal reference modulating signal using the scale-based linearly combined scaling and synthesis scaling functions. The wavelet modulation technique generates the inverter switching signals without the need for a carrier signal, precalculated switching times, or a linear range of modulation indices simplifies the implementation process. As a result, the wavelet modulation technique minimizes the memory requirements and lowers the computational burden along with insensitivity to the implementation platform.

The new wavelet modulated inverter is systematically presented in this book. The introductory chapter briefly presents the fundamental topologies and operation of power inverters. The second chapter describes wavelet basis functions and sampling theory with particular reference to the switching model of inverters. Chapter 3 outlines the connection between the non-uniform sampling theorem and wavelet functions to develop an ideal sampling-reconstruction process to operate an inverter for obtaining its optimal performances. The scale-based linearly combined basis functions are developed in Chapter 4 in order to successfully operate single-phase wavelet modulated inverters. Chapter 4 also describes the development of the non-dyadic type multiresolution analysis, which are responsible for sampling and reconstruction of three continuous time reference modulating signals for three-phase inverters. The performances of single-phase wavelet modulated inverters for static, dynamic, and non-linear loads are presented in Chapter 5, while Chapter 6 contains the simulation and experimental performances of three-phase wavelet modulated voltage source inverters for different loads at various operating conditions.

This book presents the latest technology in the advancing power electronics field. It serves as a new reference for academic researchers, curious engineers, and other professionals. The book also provides pertinent technical background for a text or reading book for a graduate-level course in advanced power electronics.

The authors acknowledge the contributions of graduate students, post-doctoral fellows, faculty, and staff at Memorial University of Newfoundland and Tokyo University of Science.

*St. John's, Newfoundland, November 2010*          *S. A. Saleh and M. A. Rahman*

# List of Symbols

| | |
|---|---|
| $\oplus$ | The direct sum of two disjoint linear subspaces |
| $\delta(t)$ | The Dirac delta function |
| $\gamma d$ | The scale-time interval factor for the sample group $d$ |
| $\widehat{ac}(\omega)$ | The DTFT of the autocorrelation sequence $ac[k]$ |
| $\lambda_{ad}(t)$ | An interpolating function over the sample group $d$ of phase $A$ |
| $\lambda_{bd}(t)$ | An interpolating function over the sample group $d$ of phase $B$ |
| $\lambda_{cd}(t)$ | An interpolating function over the sample group $d$ of phase $C$ |
| $\lambda_d(t)$ | The interpolating function over the sample group $d$ |
| $\left(\tilde{\psi}_\varphi\right)_j(t)$ | The dual scale-based linearly combined wavelet function at scale $j$ |
| $(\tilde{\varphi}_a(t))$ | The synthesis scale-based linearly combined scaling function to reconstruct $S_{Ma}(t)$ (the reference-modulating signal for phase A) |
| $(\tilde{\varphi}_b(t))$ | The synthesis scale-based linearly combined scaling function to reconstruct $S_{Mb}(t)$ (the reference-modulating signal for phase $B$) |
| $(\tilde{\varphi}_c(t))$ | The synthesis scale-based linearly combined scaling function to reconstruct $S_{Mc}(t)$ (the reference-modulating signal for phase $C$) |
| $(\varphi_a(t))$ | The scale-based linearly combined scaling function to sample $S_{Ma}(t)$ |
| $(\varphi_b(t))$ | The scale-based linearly combined scaling function to sample $S_{Mb}(t)$ |
| $(\varphi_c(t))$ | The scale-based linearly combined scaling function to sample $S_{Mc}(t)$ |
| $(L_a x_c)(t)$ | The orthogonal projection mapping operator on $x_c(t)$ |
| $(O_a x_c)(t)$ | The sampling operator on $x_c(t)$ |
| $\langle , \rangle$ | The inner product operation |
| $\{(t_{dp})_a\}$ | The set of time locations for the non-uniform samples over one cycle of $S_{Ma}(t)$ |
| $\{(t_{dp})_b\}$ | The set of time locations for the non-uniform samples over one cycle of $S_{Mb}(t)$ |
| $\{(t_{dp})_c\}$ | The set of time locations for the non-uniform samples over one cycle of $S_{Mc}(t)$ |
| $\mathbb{R}$ | The set of real numbers |

| | |
|---|---|
| $\mathbb{Z}$ | The set of integer numbers |
| $v_o(t)$ | The time function of the inverter output |
| $\omega_c$ | The frequency of the carrier signal in *rad/sec.* |
| $\omega_m$ | The frequency of the reference-modulating signal *rad/sec.* |
| $\phi_{j,k}(t)$ | A scaling function at scale $j$ and translation $k$ |
| $\psi_\varphi(t)_j$ | The scale-based linearly combined wavelet function at scale $j$ |
| $\psi_H(t)$ | The Haar wavelet function |
| $\psi_{j,k}(t)$ | A wavelet function at scale $j$ and translation $k$ |
| $\theta_c$ | An arbitrary phase shift of the carrier signal |
| $\theta_m$ | An arbitrary phase shift of the reference-modulating signal |
| $\tilde{\phi}_{j,k}(t)$ | The dual scaling function at scale $j$ and translation $k$ |
| $\tilde{\psi}_{j,k}(t)$ | The dual wavelet function at scale $j$ and translation $k$ |
| $\tilde{\varphi}_j(t)$ | The dual scale-based linearly-combined scaling function at scale $j$ |
| $ac[k]$ | An autocorrelation sequence |
| $C_{mn}$ | Complex coefficients of Fourier series |
| $f_c$ | The carrier signal frequency in *Hz* |
| $f_m$ | The frequency of the inverter output $v_o(t)$ |
| $f_m$ | The reference-modulating signal frequency in *Hz* |
| $G_d(t)$ | The Lagrange interpolating function over the sample group $d$ |
| $H$ | The Hilbert space |
| $h_\varphi[k]$ | The refinement filter associated with $\varphi(t)$ |
| $H_{sn}(f)$ | The frequency response of a synthesis filter |
| $I_L$ | The load current |
| $I_L(f)$ | The spectrum of the load current |
| $I_m$ | The motor current |
| $j$ | The scale (dilation) |
| $l^1$ | The space of absolutely convergent functions |
| $L^2$ | Space of square-integrable functions |
| $ON_{\tilde{\varphi}}$ | The duration of the ON switching pulse created by $\tilde{\varphi}_{j,k}(t)$ |
| $P_a(t)$ | The train of impulses to sample the reference-modulating signal for phase $A$ |
| $P_b(t)$ | The train of impulses to sample the reference-modulating signal for phase $B$ |
| $P_c(t)$ | The train of impulses to sample the reference-modulating signal for phase $C$ |
| $Q_i(t)$ | The switching action of the $i_{th}$ switching element |
| $R_C$ | The equivalent resistance of a semiconductor switching element |
| $S_{dM}[n]$ | The non-uniform recurrent sampled version of $S_M(t)$ (the reference-modulating signal) |
| $S_{dM}[n]_a$ | The non-uniform recurrent sampled version of $S_{Ma}(t)$ (the reference-modulating signal for phase $A$) |

| | |
|---|---|
| $S_{dM}[n]_b$ | The non-uniform recurrent sampled version of $S_{Mb}(t)$ (the reference-modulating signal for phase $B$) |
| $S_{dM}[n]_c$ | The non-uniform recurrent sampled version of $S_{Mc}(t)$ (the reference-modulating signal for phase $C$) |
| $S_{Ma}(t)$ | The reference-modulating signal for phase $A$ |
| $S_{Mb}(t)$ | The reference-modulating signal for phase $B$ |
| $S_{Mc}(t)$ | The reference-modulating signal for phase $C$ |
| $S_{SCF}[n]$ | The discrete signal created by falling portions of $S_{SC}(t)$ |
| $S_{SCR}[n]$ | The discrete signal created by rising portions of $S_{SC}(t)$ |
| $Sa_{s;a}(t)$ | The Shannon sampling function |
| $t_{1j}$ | Time location of the first sample in the sample group $d$ |
| $t_{2j}$ | Time location of the second sample in the sample group $d$ |
| $T_m$ | The period of the inverter output $v_o(t)$ |
| $T_m$ | The period of the sampled continuous-time (CT) signal |
| $T_Q$ | The Nyquist period |
| $t_{sp}$ | The time interval separating time intervals of two successive sample groups |
| $V_{AB}$ | The line-to-line voltage of phase $A$ |
| $V_{AN}$ | The line-to-neutral voltage of phase $A$ |
| $V_j$ | A space spanned by scaling basis functions at scale $j$ |
| $V_o(f)$ | The spectrum of the output voltage |
| $V_q(t)$ | The time-function of the inverter leg $q$ voltage |
| $W_j$ | A space spanned by wavelet basis functions at scale $j$ |
| $X_c(\Omega)$ | The continuous time Fourier transform of $x_c(t)$ |
| $z_b$ | The shift in the scaling function $(\varphi_b(t))_j$ |
| $z_c$ | The shift in the scaling function $(\varphi_c(t))_j$ |
| $\mathcal{B}^2_{2\Omega_o}(\mathbb{R})$ | The space of all signals band limited to $\Omega_o$ |
| $\mathcal{C}_b[0,T_s]$ | The space of continuous and bounded functions on $[0, T_s]$ |
| $\mathcal{C}_n$ | The space of complex functions |
| $\mathcal{D}$ | The number of sample groups created by $S_{SC}(t)$ over one cycle of $S_M(t)$ |
| $\mathcal{D}_\varphi$ | The number of sample groups created by $\varphi_{j,k}(t)$ over one period of $x_c(t)$ |
| $\mathbb{C}$ | The set of complex numbers |

# List of Abbreviations

| | |
|---|---|
| $1\phi$ | Single-phase |
| $3\phi$ | Three-phase |
| clos | The closure operation |
| CS | Current-source |
| CT | Continuous-time |
| CTFT | Continuous-time Fourier transform |
| DM | Delta modulation |
| DO/P | Digital output port of the ds1102 board |
| DPWM | Direct pulse-width modulation |
| DSP | Digital signal processing |
| DT | Discrete-time |
| DTFT | Discrete-time Fourier transform |
| FT | The Fourier Transform |
| $hp$ | Horse power; $1\,hp = 746$ watts |
| HPF | High-pass filter |
| $I_1$ | The fundamental component of the current |
| IGBT | Insulated gate bipolar transistor |
| inf | The infimum: the greatest lower bound of a set $S$ |
| LPF | Low-pass filter |
| MRA | Multiresolution analysis |
| NSPWM | Naturally sampled pulse-width modulation |
| PWM | Pulse-width modulation |
| $Q_i$ | The $i^{th}$ switching element of an inverter |
| RMS | Root-mean square |
| RPM | Revolution per minute |
| RPWM | Random PWM |
| RSPWM | Regularly sampled PWM |
| SBM | Sampling-based model |
| SHE | Specific harmonic elimination |
| SPWM | Sinusoidal pulse-width modulation |
| sup | The supremum: the least upper bound of a set $S$ |
| SVM | Space-vector modulation |
| THD | Total harmonic distortion |
| UPS | Uninterruptible power supplies |
| $V_1$ | The fundamental component of the voltage |
| VS | Voltage-source |
| $V_{DC}$ | Inverter input dc voltage |
| WKS | Whittaker-Koteln̆ikov-Shannon |
| WM | Wavelet modulated |
| $Z_L$ | Load impedance |

# Chapter 1

# Introduction to Power Inverters

Dc-ac power converters are part of the general power electronic converters family and are designed and operated to convert electric energy from one stage voltage, current, and/or frequency to another. Dc-ac converters, as with other power electronic converters, are composed of groups of semiconductor switching elements and are operated in a particular sequential manner to produce outputs with predefined specifications (voltage, current, and/or frequency). In general, power electronic converters operate by switching their elements in either full ON or full OFF modes in a sequential periodic manner to meet sets of predefined conditions on the output stage. The dc-ac converter is usually called the inverter. Hence, the term *dc-ac converter* will be replaced by simply *the inverter* in this book.

Inverters have gained ever-increasing popularity in a wide range of industrial applications, including ac motor drives, control systems, power supplies, uninterruptible power supply (UPS) systems, power quality, power systems, and renewable energy utilization. The majority of these applications utilize sets of conditions to ensure acceptable levels of power quality. Such sets of conditions for power quality have become standards for allowable levels of harmonic generation and distortion in inverter outputs. As a consequence, inverter performance been pushed toward modes of operation that has meet the standards imposed for output quality.

## 1.1 FUNDAMENTAL INVERTER TOPOLOGIES

Inverters are composed of groups of semiconductor switching elements such as the insulated gate bipolar transistor (IGBT), metal-oxide-semiconductor field–effect transistor (MOSFET), and bipolar junction transistor (BJT). These switching elements can be represented by a resistance $R_C$ such that:

$$OFF \Rightarrow R_C \to \infty \tag{1.1}$$

$$ON \Rightarrow R_C \to 0 \tag{1.2}$$

*An Introduction to Wavelet Modulated Inverters*, by S.A. Saleh and M. Azizur Rahman

The connection of switching elements for constructing inverters must meet the rules of basic circuit theory, in particular, transient operating conditions. Such conditions impose the following two basic constraints on the operation of switching elements in inverters:

1. A switching element is not to be connected in series with a current source (an inductive element). This is required to ensure a finite voltage across the switching element during status change (i.e., ON to OFF or OFF to ON). Generally, any combination of ON and OFF switching elements should never open-circuit an inductor.

2. A switching element is not to be connected in parallel with a voltage source (a capacitive element). This ensures a finite current through the switching element during status change. The general rule, therefore, is that any combination of ON and OFF switching elements should never short-circuit a capacitor.

Inverters can be voltage-source (VS) inverters (e.g. a dc voltage is used on the input side), or current-source (CS) inverters (e.g. a dc current is used on the input side). In general, inverters are categorized into two main types:

1. Single-phase ($1\phi$) inverters
2. Three-phase ($3\phi$) inverters

### 1.1.1   Single-Phase ($1\phi$) Inverters

There are two common topologies for single-phase ($1\phi$) inverters known as the half-bridge or the center-tapped and the full-bridge or the $H$-bridge topologies. The first topology of $1\phi$ inverters is the half-bridge or the center-tapped inverters. Figure 1.1 shows a schematic diagram for the half-bridge $1\phi$ inverter along with the switching and output voltage waveforms. The basic operation of the half-bridge inverter can be represented by defining sets of switching functions so that:

$$OFF \Rightarrow 0 \tag{1.3}$$

$$ON \Rightarrow 1 \tag{1.4}$$

These sets of switching functions are capable of defining the output of the inverter as a function of the switching actions. The output voltage waveform shown in Figure 1.1 can be described by the following equation as:

$$v_0(t) = \frac{V_{DC}}{2}(Q_1(t) - Q_2(t)) \tag{1.5}$$

where $Q_1(t)$ and $Q_2(t)$ are switching actions of the switching elements $Q_1$ and $Q_2$, respectively. Because $Q_1$ and $Q_2$ cannot be ON at the same time, as well as for continuous output current, the relation between $Q_1(t)$ and $Q_2(t)$ can be stated as:

$$Q_1(t) + Q_2(t) = 1 \tag{1.6}$$

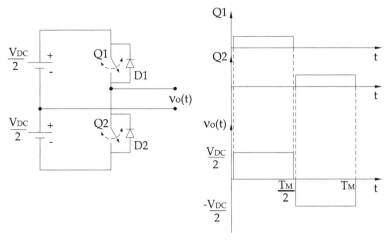

**Figure 1.1**  Schematic diagram of the half-bridge $1\phi$ inverter along with switching and output waveforms.

This provides the output voltage as a function of the switching actions as:

$$v_0(t) = \frac{V_{DC}}{2}(2Q_1(t) - 1) \tag{1.7}$$

The output voltage of the half-bridge inverter $v_o(t)$ has a root mean square (RMS) value given as:

$$(v_0)_{RMS} = \left( \frac{2}{T_M} \int_0^{\frac{T_M}{2}} \frac{V_{DC}^2}{4} dt \right)^{\frac{1}{2}} = \frac{V_{DC}}{2} \tag{1.8}$$

The second topology of $1\phi$ inverters is the full-bridge or the $H$-bridge that has two single-phase legs of the same type. This inverter topology is preferred in high power applications due to the fact that the output voltage is twice that of the half-bridge inverter. Also, the current flowing in each switch is half the current that flows in each switch of the half-bridge inverter. Figure 1.2 shows a schematic diagram for the full-bridge or the $H$-bridge inverter along with the switching and the output voltage waveforms. The output voltage waveform shown in Figure 1.2 can be described as:

$$V_A(t) = \frac{V_{DC}}{2}(Q_1(t) - Q_4(t)) \tag{1.9}$$

$$V_B(t) = \frac{V_{DC}}{2}(Q_3(t) - Q_2(t)) \tag{1.10}$$

where $Q_1(t)$, $Q_2(t)$, $Q_3(t)$, and $Q_4(t)$ are switching actions of the switching elements $Q_1$, $Q_2$, $Q_3$, and $Q_4$, respectively. Since $Q_1$ and $Q_4$ or $Q_3$ and $Q_2$ cannot be ON at the same time as continuous output current, the relation between $Q_1(t)$, $Q_2(t)$, $Q_3(t)$, and $Q_4(t)$ can be stated as:

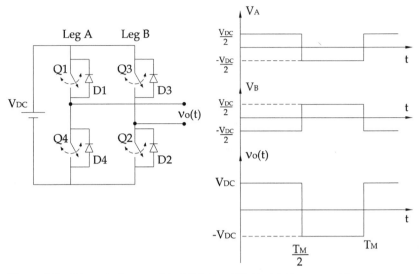

**Figure 1.2** Schematic diagram of the full-bridge ($H$-bridge) $1\phi$ inverter and the voltage waveforms. The output voltage is $v_o(t) = V_A(t) - V_B(t)$.

$$Q_1(t) + Q_4(t) = 1 \qquad (1.11)$$

$$Q_3(t) + Q_2(t) = 1 \qquad (1.12)$$

This can provide the output voltages as functions of the switching actions as:

$$V_A(t) = \frac{V_{DC}}{2}(2Q_1(t) - 1) \qquad (1.13)$$

$$V_B(t) = \frac{V_{DC}}{2}(2Q_3(t) - 1) \qquad (1.14)$$

The output voltage of the full-bridge inverter $v_o(t)$ can be expressed as:

$$v_o(t) = V_{DC}(Q_1(t) - Q_3(t)) \qquad (1.15)$$

The output voltage of the full-bridge inverter $v_o(t)$ has an RMS value given as:

$$(v_o)_{RMS} = \left( \frac{2}{T_M} \int_0^{\frac{T_M}{2}} V_{DC}^2 \, dt \right)^{\frac{1}{2}} = V_{DC} \qquad (1.16)$$

## 1.1.2 Three-Phase ($3\phi$) Inverters

Three-phase inverters are used for variable-frequency drive applications and for high power applications such as HVDC power transmission. The conventional three-phase inverter consists of three single-phase legs, each connected to one phase on the output side. The fundamental operation of the three legs is coordinated so that

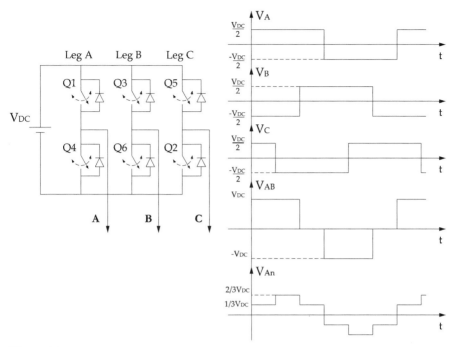

**Figure 1.3**    Schematic diagrams of the three-phase ($3\phi$) inverter and the output voltage waveforms.

one switch operates at each 60-degree point of the fundamental output waveform. This creates a line-to-line output waveform that has six steps. The six-step waveform has a zero-voltage step between the positive and negative sections of the square-wave such that the harmonics that are multiples of three are eliminated. Note that switching elements of $3\phi$ inverters can be operated in the 180-degree-conduction and the 120-degree-conduction. Figure 1.3 shows the schematic diagram of the $3\phi$ inverter along with the switching and output waveforms for the 180-degree-conduction operation. The output voltage waveforms across each leg shown in Figure 1.3 can be described as:

$$V_A(t) = \frac{V_{DC}}{2}(Q_1(t) - Q_4(t)) \tag{1.17}$$

$$V_B(t) = \frac{V_{DC}}{2}(Q_3(t) - Q_6(t)) \tag{1.18}$$

$$V_C(t) = \frac{V_{DC}}{2}(Q_5(t) - Q_2(t)) \tag{1.19}$$

where $Q_1(t)$, $Q_2(t)$, $Q_3(t)$, $Q_4(t)$, $Q_5(t)$, and $Q_6(t)$ are switching actions of the switching elements $Q_1$, $Q_2$, $Q_3$, $Q_4$, $Q_5$, and $Q_6$, respectively. Since $Q_1$ and $Q_4$, $Q_3$, and $Q_6$, and

$Q_5$ and $Q_2$ cannot be ON at the same time as continuous output current, the following relations can be stated:

$$Q_1(t)+Q_4(t)=1 \tag{1.20}$$

$$Q_3(t)+Q_6(t)=1 \tag{1.21}$$

$$Q_5(t)+Q_2(t)=1 \tag{1.22}$$

This can provide the leg voltages as functions of the switching actions as:

$$V_A(t)=\frac{V_{DC}}{2}(2Q_1(t)-1) \tag{1.23}$$

$$V_B(t)=\frac{V_{DC}}{2}(2Q_3(t)-1) \tag{1.24}$$

$$V_C(t)=\frac{V_{DC}}{2}(2Q_5(t)-1) \tag{1.25}$$

The output line-to-line voltages of the $3\phi$ inverter can be expressed as:

$$v_{AB}(t)=V_A(t)-V_B(t)=V_{DC}(Q_1(t)-Q_3(t)) \tag{1.26}$$

$$v_{BC}(t)=V_B(t)-V_C(t)=V_{DC}(Q_3(t)-Q_5(t)) \tag{1.27}$$

$$v_{CA}(t)=V_C(t)-V_A(t)=V_{DC}(Q_5(t)-Q_1(t)) \tag{1.28}$$

The inverter topologies are very popular for widespread applications in different industrial sectors. The demand for utilizing inverters in industrial applications with high power has pushed toward developing new inverter topologies with high power ratings. Such topologies are characterized by power levels, output quality, switching elements, and diversity of topological designs and structures. These inverter topologies have created a new family of inverters called multi-level inverters.

## 1.2  MULTILEVEL INVERTER TOPOLOGIES

The conventional inverter topologies operation is based on switching each leg to positive, zero, and negative voltage levels (the same applies for current source inverters (CSI)). Such inverter topology can be limited to the current as well as voltage ratings of the switching elements used for high power applications. One of the latest approaches to overcome such inverter limitations is through increasing the number of voltage levels (or current levels in CSI) on the output side of the inverter. This has lead to the concept of multi-level inverters. It is worth mentioning that the concept of multilevel inverters is applicable for both $1\phi$ and $3\phi$ inverters, however, the majority of multilevel inverters are designed as $3\phi$ ones.

## 1.2.1  Neutral-Point Clamped Multilevel Inverter

The first multilevel inverter was the $3\phi$ neutral-point clamped (NPC) inverter that has a zero-voltage center point, which can be switched to output phases. The switching of the zero-voltage points to the output phases made it possible to switch each inverter leg to three levels of voltage. Figure 1.4 shows a schematic diagram of a $3\phi$ three-level VS NPC inverter. One of the advantages of the NPC multilevel inverters is that they have twice the number of the switching elements as the six-pulse $3\phi$ inverters, where each switching element blocks only half the dc bus voltage. Also, the increase in the number of switching elements guarantees a reduced switch utilization.

The purpose of producing different levels of voltage is to sequentially activate only two switching elements in each leg. For the three-level NPC inverter shown in Figure 1.4, the output voltages can be expressed as functions of the switching actions:

$$V_{AB}(t) = V_A(t) - V_B(t) = V_{DC}(Q_{a1}(t) - Q_{a3}(t) - Q_{b1}(t) + Q_{b3}(t)) \qquad (1.29)$$

$$V_{BC}(t) = V_B(t) - V_C(t) = V_{DC}(Q_{b1}(t) - Q_{b3}(t) - Q_{c1}(t) + Q_{c3}(t)) \qquad (1.30)$$

$$V_{CA}(t) = V_C(t) - V_A(t) = V_{DC}(Q_{c1}(t) - Q_{c3}(t) - Q_{a1}(t) + Q_{a3}(t)) \qquad (1.31)$$

The aforementioned equations for the output line-to-line voltages indicate that each one of the three voltages can have $\pm 2V_{DC}$, $\pm V_{DC}$, or zero. Figure 1.5 shows the output voltage waveforms for the leg A voltage, line-to-line voltage, and line-to-neutral voltage.

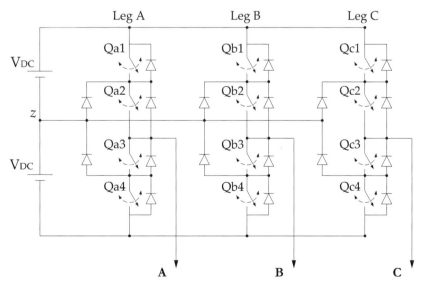

**Figure 1.4**  Schematic diagram of a $3\phi$ three-level voltage source (VS) neutral-point clamped (NPC) inverter.

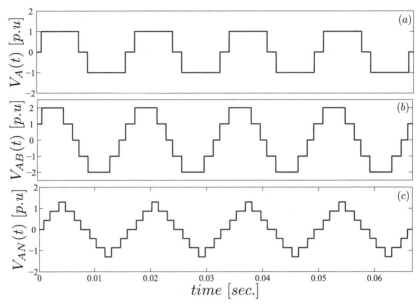

**Figure 1.5**    The three-level NPC inverter output voltages: (a) the NPC inverter per-unit leg A output voltage $V_A$, (b) the NPC inverter per-unit line-to-line output voltage $V_{AB}$, and (c) the NPC inverter per-unit line-to-neutral output voltage $V_{AN}$. The voltage base value is the dc voltage $V_{DC}$.

## 1.2.2  Diode-Clamped Multilevel Inverter

Diode-clamped multilevel (DCM) inverters extend the concept of several levels of the output voltage achieved using the NPC inverters. These inverters use diodes to limit the voltage stress across the switching elements so that different levels of voltage can be produced on the inverter output. The basic structure of the DCM inverter utilizes switching elements connected in series with their respective main diodes. All are connected in parallel with the clamping diodes. In each leg of the inverter, the forward voltage across each switching element is clamped by the connection of diodes between switching elements and the nodes. Figure 1.6 shows a schematic diagram of a $3\phi$ four-level VS DCM inverter.

Note that each active switching element blocks one level of the dc bus voltage. However, the clamping diodes may block a voltage higher than the one blocked by the switching element, indicating that their ratings for reverse voltage blocking need to be set adequately.

## 1.2.3  Capacitor-Clamped Multilevel Inverter

The capacitor-clamped multilevel (CCM) inverter offers the same advantages as both NPC and DCM inverters in the ability to produce output voltages with different levels. However, the CCM inverter offers additional operational advantages that

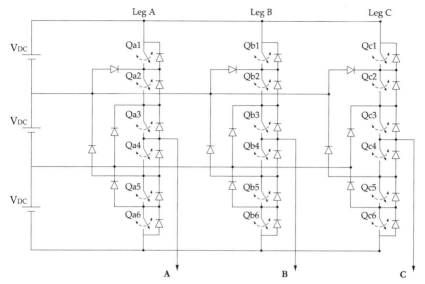

**Figure 1.6** Schematic diagram of a $3\phi$ four-level voltage source (VS) diode-clamped multilevel (DCM) inverter.

include lower cost and higher efficiency due to the fact that fewer diodes are used. The operation of this inverter topology can be simply described as follows: the capacitors in each leg are charged to different voltage levels. Figure 1.7 shows a schematic diagram of a $3\phi$ four-level VS CCM inverter.

A combination of ON switching elements allows different voltages from different capacitors to synthesize the output voltage. The voltage waveforms shown in Figure 1.5 can be obtained using a three-level VS capacitor-clamped multilevel inverter.

### 1.2.4 Cascaded *H*-Bridge Multilevel Inverter

The cascaded single-phase *H*-bridge multilevel inverter uses different single-phase *H*-bridge ones, each with an independent dc voltage source. The different legs of a cascaded *H*-bridge multilevel inverter can have different levels of voltage that are switched by the individual $1\phi$ *H*-bridge inverters, where three levels of voltage ($V_{DC}$, 0, and $-V_{DC}$) can be obtained. The ac outputs of the $1\phi$ *H*-bridge inverters in each leg are connected in series such that the synthesized voltage waveform is the sum of the *H*-bridge inverters' outputs. Figure 1.8 shows a schematic diagram of a $3\phi$ five-level VS cascaded single-phase *H*-bridge multilevel inverter.

The general rule for the number of levels in the output phase-to-neutral voltage VS cascaded single-phase *H*-bridge multilevel inverter is stated as:

$$m = 2s + 1 \tag{1.32}$$

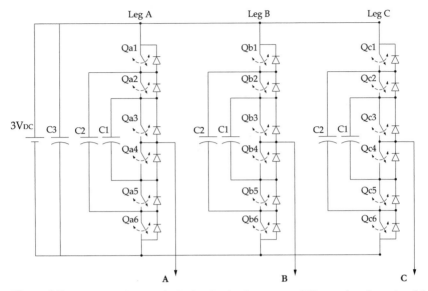

**Figure 1.7**    Schematic diagram of a 3$\phi$ four-level voltage source (VS) capacitor-clamped multilevel (CCM) inverter.

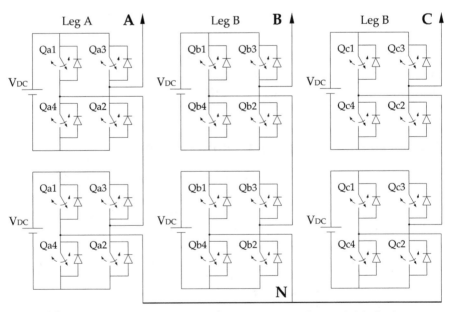

**Figure 1.8**    Schematic diagram of a 3$\phi$ four-level voltage source (VS) cascaded single-phase *H*-bridge multilevel inverter.

where $m$ is the number of voltage levels and $s$ is the number of dc voltage supplies. The main advantages of the cascaded single-phase $H$-bridge multilevel inverter are its ability to offer improved regulation of the dc bus voltage and its structure, which is modular to simplify control as well as maintenance.

There are other topologies for multilevel inverters that are mostly based on hybrid combinations of the aforementioned multilevel inverters. Furthermore, these hybrid multilevel inverters are designed and operated to meet specific conditions required by certain applications. Among these hybrid multilevel inverters are the generalized multilevel topology, the mixed-level hybrid multilevel topology, soft-switched multilevel topology, and the back-to-back diode-clamped multilevel topology.

## 1.3  FUNDAMENTAL INVERTER SWITCHING

The principle of inverter operation is thoroughly discussed in the literature. Inverter operation coordinates the states of all switching elements and ensures their compliance with fundamental conditions for switching circuits. These conditions are required to avoid creating short circuits across the dc supply and to provide each switching element with the required time for changing its status from ON to OFF or OFF to ON. Adherence to these conditions by certain sequential switching method produces ac outputs. However, outputs of these inverters contain different frequency components in addition to the desired fundamental frequency component. Such frequency components can create undesired features in the ac outputs as well as various levels of operational imperfections.

Concerns about inverter output quality, which reflects the amount of output energy distributed in the harmonic frequencies, have been a challenging issue since the first application of inverters as power conditioners. subsequently, operating inverters employing various switching strategies to reduce the output harmonic generation and distortion have been a topic for intensive research. The main objectives of such research are to develop and test switching techniques capable of operating inverters to produce outputs with high quality over wide ranges of loading conditions. These objectives can be interpreted as concentrating as much energy of the inverter output as possible in the desired frequency component. Two fundamental strategies have been developed to operate both $1\phi$ and $3\phi$ inverters that are the square-wave and multi-switching strategies. Figure 1.9 shows waveforms for an inverter output operated using both the square-wave and multiswitching strategies. The spectra of the inverter outputs of Figure 1.9 are shown in Figure 1.10. The multiswitching strategy has been found effective in reducing the output harmonic content and, as a result, is capable of improving the quality of inverter outputs. The multiswitching strategy is based on activating a group of switching elements ON and OFF with a rate higher than the desired output frequency. The switching technique of changing the status of a switching element from ON to OFF or OFF to ON is known as the modulation technique. There are several modulation techniques that

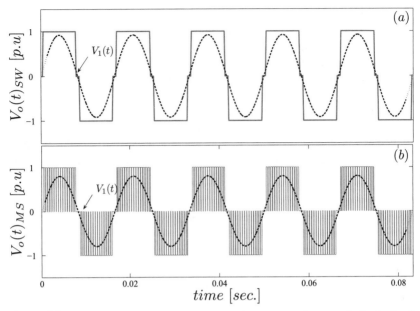

**Figure 1.9**   Strategies for inverter operation: (a) the inverter per-unit output obtained using the square-wave (SW) strategy and (b) the inverter per-unit output obtained using the multiswitching (MS) strategy. $V_1(t)$ the time function of the fundamental component of $V_o(t)$.

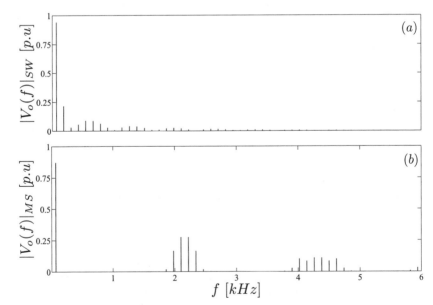

**Figure 1.10**   Strategies for inverter operation: (a) the per-unit magnitude spectrum of the inverter output obtained using the square-wave (SW) strategy and (b) the per-unit magnitude spectrum of the inverter output obtained using the multiswitching (MS) strategy.

have been developed and tested for operating inverters to achieve the aforementioned objectives by meeting the following conditions:

- wide linear modulation range
- minimum switching loss and improved overall inverter efficiency
- high output quality evaluated through the total harmonic distortion (THD) factors of inverter outputs
- high magnitude of the output fundamental frequency component
- simple implementation for practical applications
- low computation time

Various approaches have been used to develop new modulation techniques that can improve the performance of both $1\phi$ and $3\phi$ inverters. These approaches have provided continuous advancements in both solid-state technology and digital systems. One of the remarkable new technologies in operating inverters is the ability to realize high switching frequencies in stable and reliable manners. Also, with these advancements and the developments in micro-processor technology, modulation techniques that require complex computations have become implementable for both testing and operation levels. Application of inverters in high power applications has made switching losses, switching capabilities, and inverter efficiency critical issues that must be taken into account in performance evaluation. The fundamental approaches to inverter modulation are as follows:

- Carrier-based pulse-width modulation (PWM) generates switching pulses by comparing a high-frequency carrier signal with a low-frequency sinusoidal modulating signal. The PWM inverters offers linear modulation, improved quality relative to the single-pulse (square-wave) switching strategy, and were the first implementation of multiswitched inverters. The carrier-based PWM technique and its subsequent improvements are very popular in a wide range of applications. As the research to improve inverter performance continued, the non-sinusoidal carrier-based PWM technique was introduced. This technique was developed to extend the linear modulation range and to improve the magnitude of the output fundamental frequency component (the desired frequency component). The nonsinusoidal PWM was further improved and defined as space-vector modulation (SVM), which is only used for $3\phi$ inverters. It is worth mentioning that PWM inverters are very popular and are used extensively in various industrial applications.
- Selected harmonic elimination (SHE) calculates the switching angles to determine the locations and widths of the switching pulses. SHE inverters offer the possibility of eliminating certain frequency components (harmonics) present in the inverter output.
- Random PWM (RPWM) concentrates energy in the inverter output in a narrow frequency band by randomizing the frequency of the carrier-signal. RPWM inverters provided further improvements on the inverter performance

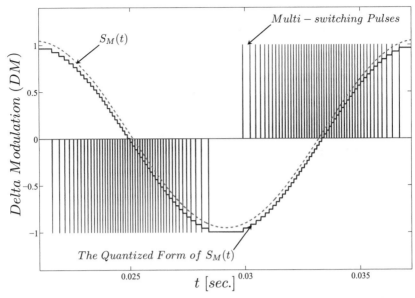

**Figure 1.11**    Quantization of the reference-modulating signal and the generated pulses as segment heights change.

through the effective reduction of the energy distributed in the harmonic frequency bands, while concentrating most of the energy of the inverter output in the desired frequency component.

- The delta modulation (DM) technique approximates the reference-modulating signal by sinusoid piece-wise linear segments. Each segment is compared to the reference-modulating signal to determine the increase or decrease in its relative amplitude. Only the change in amplitude is considered for changing the state of the modulated signal. In the DM technique, pulse widths are not modulated but rather have constant widths. The correct terminology for such a modulation technique is pulse density modulation (PDM) or pulse frequency modulation (PFM). The DM technique is the simplest method for quantizing analog signals into digital sequences of data with significant accuracy. This accuracy can be achieved by using switching frequencies much higher than the frequency of the reference-modulating signal. Figure 1.11 shows simple DM multi-switching signals generated by quantizing a sinusoidal reference-modulating with a sampling interval of 0.00001 second, while the quantizer period is around 0.0375 second.

Other approaches are reported in the literature that have been used to develop inverter switching techniques, among which are the direct pulse-width modulation technique (DPWM), the optimal PWM, and others.

## 1.4  HARMONIC DISTORTION

Harmonics are defined as high-frequency components of currents or voltages in electric systems. These frequency components are integer multiples of the fundamental frequency. Harmonic contents of any power supply can create different problems, including power losses, mechanical resonances due to magnetic fields generated by such frequency components, overheating of motors, and high currents in neutral paths. The magnitudes of harmonic contents can cause harmonic distortion, which reflects the amount of energy distributed in the harmonic frequencies relative to that contained in the fundamental frequency component. The introduction of various algorithms and procedures to realize different inverter modulation techniques has created a significant demand for performance comparison. Furthermore, with each new modulation technique, a simple performance criterion is required to highlight the advantages of such a new technique as well as to demonstrate the capabilities of overcoming difficulties faced by other techniques.

The inverter output $v_o(t)$ is an aperiodic function that has a finite period $T_m$. The function $v_o(t)$ has a root mean square (RMS) that is given as:

$$(V_o)_{rms} = \left( \frac{1}{T_m} \int_0^{T_m} v_o^2(t)\, dt \right)^{\frac{1}{2}} \tag{1.33}$$

Also, the periodic nature of $v_o(t)$ allows its representation using the Fourier series:

$$v_o(t) = \frac{a_0}{2} \sum_{n=1}^{\infty} a_n \cos(2n\pi f_m t) + b_n \sin(2n\pi f_m t) \tag{1.34}$$

where $f_m = \dfrac{1}{T_m}$ is the frequency of the inverter output $v_o(t)$. The Fourier series coefficients $\{a_n\}$ and $\{b_n\}$ are given by:

$$a_n = \frac{1}{T_m} \int_0^{T_m} v_o(t) \cos(2n\pi f_m t)\, dt \quad n = 0, 1, 2 \ldots \infty \; and \;\; 0 < t < T_m \tag{1.35}$$

$$b_n = \frac{1}{T_m} \int_0^{T_m} v_o(t) \sin(2n\pi f_m t)\, dt \quad n = 1, 2, 3 \ldots \infty \; and \;\; 0 < t < T_m \tag{1.36}$$

The inverter output $v_o(t)$ is an odd function, which sets the coefficient as $\{an\} = 0$. Moreover, due to the quarter-cycle symmetry of $v_o(t)$, all the even frequency components will have $\{b_n\}_{n=2,4,6\ldots}$ evaluated with 0. These two features of the inverter output $v_o(t)$ simplify the Fourier series as:

$$v_o(t) = \sum_{n=1,3,5\ldots}^{\infty} b_n \sin(2n\pi f_m t) \tag{1.37}$$

Substituting Fourier series expansion into equation (1.33) yields:

$$(V_o)_{rms} = \left( \frac{1}{T_m} \int_0^{T_m} \sum_{n=1,3,5\ldots}^{\infty} b_n^2 \sin^2(2n\pi f_m t)\, dt \right)^{\frac{1}{2}} \tag{1.38}$$

$$(V_o)_{rms} = \left( \frac{1}{T_m} \int_0^{T_m} \sum_{n=1,3,5...}^{\infty} \frac{b_n^2}{2} (1 - \cos(4n\pi f_m t)) dt \right)^{\frac{1}{2}} \tag{1.39}$$

The integration of the double-frequency terms is evaluated with 0 over the interval $T_m$. This simplifies the inverter output RMS value to:

$$(V_o)_{rms} = \left( \sum_{n=1,3,5...}^{\infty} \frac{b_n^2}{2} \right)^{\frac{1}{2}} = \left( \sum_{n=1,3,5...}^{\infty} (b_n^2)_{rms} \right)^{\frac{1}{2}} \tag{1.40}$$

As in most inverter applications, the fundamental frequency component ($n = 1$) is considered, a factor that reflects the distortion caused by other frequency components with respect to the fundamental frequency component. This factor is defined as the total harmonic distortion (THD) factor, which has the following mathematical expression:

$$THD = \frac{\sqrt{\sum_{n=3,5,...}^{\infty} V_n^2}}{V_1} \tag{1.41}$$

where $V_n = b_n$, $n = 3$, $5,....$ With the introduction of advanced multiswitching techniques, the existence of inverter output harmonic components became more complicated for the conventional Fourier series approach. In 1975, Bowes presented a double Fourier integral formula that had originally been developed for communication applications. The double Fourier integral formula approach is based on defining two independent linear time variables function as:

$$x(t) = \omega_c t + \theta_c \tag{1.42}$$

$$y(t) = \omega_m t + \theta_m \tag{1.43}$$

where $\omega_c$ is the carrier signal frequency, $\theta_c$ is an arbitrary phase shift of the carrier signal, $\omega_m$ is the modulating signal frequency, and $\theta_m$ is an arbitrary phase shift of the reference-modulating signal. A multivariable function $f(t)$ can be defined as:

$$f(t) = f(x(t), y(t)) \tag{1.44}$$

Using Fourier analysis theory, complex coefficients of Fourier series for the function $f(t)$ can be determined as:

$$C_{mn} = \frac{1}{2\pi^2} \int_{-\pi}^{\pi} \int_{-\pi}^{\pi} f(x(t), y(t)) e^{j(mx+ny)} dx dy \tag{1.45}$$

The complex coefficients $C_{mn}$ can determine the magnitude and the phase of each harmonic component present in the function $f(t)$. This analysis was used as a performance criterion for testing effective switching strategies that resulted in implementing the naturally sampled PWM scheme. In the early 1980s, the same criterion was further expanded by Bowes and Mount to successfully implement and test the regularly sampled PWM inverter using microprocessor technology.

## 1.5  SUMMARY

This chapter has presented an introduction to the fundamental structures and operation of inverters. Some sections in this introductory chapter have been devoted to a brief presentation of the popular topologies used in inverter applications along with their structure and basis of operation. Furthermore, the two basic inverter switching methods have been discussed, along with their general waveforms. The last part of this chapter has focused on the Fourier analysis methods that are used to model the distortions in inverter outputs.

# Chapter 2

# Wavelets and the Sampling Theorem

## 2.1 INTRODUCTION

Processing continuous-time (CT) signals is a major step and requirement in different applications, including communication systems, control systems, power systems, power electronics, image processing, and biomedical applications. Furthermore, the search for proper techniques is one of the main challenges for accurate and realistic implementation of CT processing. There are many working techniques, including Fourier transforms, Laplace transform, and Z-transform. Such CT processing techniques work well for CT signals, which satisfy the conditions of finite energy, periodicity, stationarity, and finite number of discontinuities. In processing any signal, it is desirable and often necessary to apply a frequency-time based analysis technique in an attempt to extract all the frequency components, including the transient components present in the processed signal, and to preserve their location in time. This constraint has made it hard to process the large class of CT signals that do not satisfy the conditions of classical signal processing techniques.

In general, processing a CT signal, which involves sampling and reconstructing the CT signal, can be represented as an $N$-dimensional approximation case. Such a representation has been used to interpret the sampling theorem in the context of the wavelet-based multiresolution analysis (MRA). There are different types of wavelet basis functions that are capable of constructing MRAs, which can support sampling structures. However, such constructed MRAs are capable of supporting uniform sampling structures. The different types of available wavelet basis functions, along with the main characteristics of their associated MRAs, are provided in this chapter. Furthermore, this chapter presents a new type of wavelet basis function, capable of supporting a nonuniform recurrent sampling-reconstruction structure.

One of the most effective methods for decomposing CT signals is through an MRA that is constructed by orthogonal basis functions. In this type of signal processing, a CT signal is broken into orthogonal time-localized frequency channels (scales). The required orthogonal basis functions are generated by integer-indexed

*An Introduction to Wavelet Modulated Inverters*, by S.A. Saleh and M. Azizur Rahman
Copyright © 2011 Institute of Electrical and Electronics Engineers

translations and dyadic (*powers of* 2) dilations of a single function that is known as the scaling function ($\phi(t)$). This scaling function $\phi(t)$ when dilated to scale $j$ as $\phi_j(t)$ is orthogonal to its translations at that dilation (scale) $j$. The generated basis functions at scale $j$ span a space $V_j$ that provides an approximation to the signal in that space, which can be defined as:

$$V_j(\varphi) = \text{clos}_{L^2}\langle\{\phi_{j,k}(t)\}\rangle, \quad j = 0, 1, 2, 3, \ldots, \quad k \in \mathbb{Z} \tag{2.1}$$

where the set $\{\phi_{j,k}(t)\}$ is given as:

$$\{\phi_{j,k}(t)\} = \{\phi_1(2^j t - k)\} \, j = 0, 1, 2, \ldots k \in \mathbb{Z} \tag{2.2}$$

The notation $\text{clos}_{L^2}$ is the closure of all linear combinations of all inner products of the set $\{\phi_{j,k}(t)\}$ over $L^2$. The clos operation can be defined as follows:

*Definition:* Let $\{g_n(x)\}$ be a collection of functions that can form a linear space $\mathcal{P}$ as:

$$\mathcal{P} = \text{span}\{g_n(x)\} \tag{2.3}$$

The closure of the linear space $\mathcal{P}$ denoted by $\text{clos}_{L^2}\{\mathcal{P}\}$ can be stated as: a function $f(x) \in \text{clos}_{L^2}\{\mathcal{P}\}$ if for every $\varepsilon > 0$, there is a function $g(x) \in \mathcal{P}$ such that:

$$\|f(x) - g(x)\|_2 < \varepsilon \tag{2.4}$$

The collection of the spanned spaces $\{V_j\}$ allows the construction of a dyadic type MRA. Since scaling basis functions $\{\phi_{j,k}\}$ are generated through shifting and dilating the scaling function $\phi(t)$, the spanned scaling spaces are nested such that:

$$\ldots V_{j-1} \subset V_j \subset V_{j+1} \subset V_{j+2}. \ldots \tag{2.5}$$

Although the generated basis functions at a certain dilation (scale) $j$ ($\{\phi_{j,k}\}$) are orthogonal, they are not complete with respect to $L^2(\mathbb{R})$. A more complete set $\{\phi_{j-1,k}\}$ is also orthogonal but is twice as dense. The difference between the successive spaces $V_j$ and $V_{j-1}$ yields a difference space spanned by another set of basis functions that are known as the wavelet basis functions. That is, for each space $V_j$, there exists an orthogonal complement space $W_j$, which is spanned by the set $\{\psi_{j,k}\}$ and can be defined as:

$$W_j(\varphi) = \text{clos}_{L^2}\langle\{\psi_{j,k}(t)\}\rangle, \quad j = 1, 2, 3, \ldots, \quad k \in \mathbb{Z} \tag{2.6}$$

where $\{\psi_{j,k}(t)\}$ is defined as:

$$\{\psi_{j,k}(t)\} = \{\psi_1(2^j t - k)\} \, j = 0, 1, 2 \ldots k \in \mathbb{Z} \tag{2.7}$$

In general, a spanned space $V_j$ in an MRA can be constructed using both spaces $V_{j+1}$ and $W_{j+1}$ as:

$$V_j = V_{j+1} \oplus W_{j+1} \tag{2.8}$$

There are four major types of wavelet basis functions used in different signal and image processing applications. These types are:

1. Orthogonal wavelet basis functions
2. Semi-orthogonal wavelet basis functions
3. Bi-orthogonal wavelet basis functions
4. Shift-orthogonal wavelet basis functions

These types of wavelet basis functions are capable of spanning spaces such as $W_j$ (see equation (2.6)) and $V_j$ (see equation (2.8)). A collection of $W_j$ and $V_j$ spaces allows the construction of MRAs. However, these constructed MRAs are the dyadic type that can support uniform ideal sampling processes.

## 2.2  WAVELET BASIS FUNCTIONS

Wavelet Basis functions provide a system of coordinates in which several classes of linear operators are sparse. This system of coordinates is capable of expanding signals at different levels of resolution using coefficients in linear combinations of sets of basis functions. Temporal expansion is performed with contracted and high-frequency bases, while frequency analysis is performed with dilated and low-frequency ones. In general, wavelet basis functions can expand (decompose) signals and provide a time location for each frequency component present in such decomposed signals. These basis functions are usually related to a single function, known as the scaling function $\phi(t)$. A scaling function is a nonzero solution to a dilation equation of the form:

$$\phi(t) = \sum_{k=0}^{N} c_k \phi(2t-k) \quad c_k \in \mathbb{R}, \quad k \in \mathbb{Z} \qquad (2.9)$$

where the function $\phi(t)$ is a function that satisfies the conditions of a scaling function. The conditions required for a function to be a scaling function are described as the following:

- The function has to be a real and a continuous function with a decaying property such that:

$$\lim_{|t| \to \infty} \phi(t) = 0, \quad t \in \mathbb{R} \qquad (2.10)$$

- The integer translations of $\phi(t)$ of the form $\{\phi(t-l)\}$, $l \in \mathbb{Z}$ form an orthonormal basis for a subspace $V_0$ of $L^2(\mathbb{R})$.
- The MRA generated by $\phi(t)$ of closed subsets $\{V_m^2\}_{m \in \mathbb{Z}}$ of $L^2(\mathbb{R})$ has to satisfy:

$$0 \ldots \subset V_{-1} \subset V_0 \subset V_1 \subset V_2 \ldots \subset V_m \subset L^2(\mathbb{R}) \qquad (2.11)$$

$$x_c(t) \in V_m \Leftrightarrow x(2t) \in V_{m+1} \qquad (2.12)$$

$$\bigcap_{m \in \mathbb{Z}} V_m = 0, \ \overline{\bigcup_{m \in \mathbb{Z}} V_m} = L^2(\mathbb{R}) \qquad (2.13)$$

- Since $\phi(t) \in V_1$, there exists a sequence $\{a_k\}$ of length $N$ such that:

$$\phi(t) = \sum_{k=0}^{N-1} a_k \phi(2t - k) \tag{2.14}$$

When the aforementioned conditions are met by a function $\phi(t)$, then it is capable of generating orthonormal basis functions for constructing an MRA. In addition, there exists an associated function that is defined as the wavelet function $\psi(t)$. The wavelet function can generate a set of orthonormal basis functions for a subspace $W_0$ that is an orthogonal complement of the subspace $V_0$. Such a set of orthonormal basis functions is generated by integer translations of the wavelet function $\psi(t)$ as $\{\psi(t - l)\}$, $l \in \mathbb{Z}$.

Scaling and wavelet basis functions are stable sets of basis functions. Moreover, these basis functions are localized in time and frequency, which makes them capable of spanning complete linear spaces. In general, a set of functions is a basis if it can meet the following conditions.

- Completeness:

  A set $\{v_k(t)\}_{k \in \mathbb{Z}}$ is complete if its span is dense in a normed space $V$ such that:

$$V = \text{clos}\left(\text{span}\left(\{v_k(t)\}\right)\right) \tag{2.15}$$

  where span $(\{v_k(t)\})$ is given by:

$$\text{span}\left(\{v_k(t)\}\right) = \left\{ \sum_{k \in \mathbb{Z}} \alpha_k v_k(t) \middle| \alpha_k \in \mathbb{R} \text{ or } \mathbb{C} \right\} \tag{2.16}$$

  where $\mathbb{C}$ is the set of complex numbers.

- Linear independence:

  A set $\{v_k(t)\}_{k \in \mathbb{Z}}$ is linearly independent if and only if none of the basis functions $v_k(t)$ are contained in the linear span of the other basis functions, which can be expressed as:

$$\sum_{k \in \mathbb{Z}} \alpha_k v_k(t) = 0 \quad \text{if and only if } \alpha_k = 0, \text{ for all } k \tag{2.17}$$

- Being a basis for the Hilbert space $H$:

  A set $\{v_k(t)\}_{k \in \mathbb{Z}}$ is a basis for a Hilbert space $H$ if every function $f(t) \in H$ can be uniquely expressed as:

$$f(t) = \sum_{k \in \mathbb{Z}} c_k v_k(t) \quad c_k \in \mathbb{C} \tag{2.18}$$

The generated scaling and wavelet function can be defined as:

$$\{\phi(t - k) : k \in \mathbb{Z}\} \tag{2.19}$$

$$\{\psi(t - k) : k \in \mathbb{Z}\} \tag{2.20}$$

In general, any stable set of basis functions is capable of spanning a space $V(\phi)$ such that;

$$V(\phi) := \text{clos}_{L^2}\langle\{\phi(t-k)\}:k\in\mathbb{Z}\rangle \tag{2.21}$$

This property is valid if there exist two positive constants $A$ and $B$ called Riesz bounds for all functions $s(t) \in C_n$ such that:

$$A\|s(t)\|_2^2 \leq \sum_k |\langle s(t), \phi(t-k)\rangle|^2 \leq B\|s(t)\|_2^2 \quad \text{with } 0 < A \leq B < \infty \tag{2.22}$$

where $C_n$ is the space of complex functions and the term $\|s(t)\|_2^2$ is given by:

$$\|s(t)\|_2^2 = \int_t |s(t)|^2\,dt \tag{2.23}$$

Scaling and wavelet basis functions are special cases of Hilbert space basis functions with capabilities of constructing stable MRAs. The following subsections provide brief descriptions of conventional classes of wavelet basis functions along with main characteristics of their spanned spaces.

## 2.2.1  Orthogonal Wavelet Basis Functions

Orthogonal basis functions are the first forms of well-defined sets of basis functions used for processing signals. Fourier and Haar bases are among the earliest well-defined and popular basis functions that have been used in different signal and image processing applications. The latest advancements in the theory of signal processing have made wavelets very popular tools in many other engineering areas such as power systems, where orthogonal wavelets have been used in developing protection systems and in improving power quality.

In general, orthogonal wavelets are characterized by their ability to generate sets of orthogonal basis functions. A set of functions $\{\chi_n(t)\}_{n\in\mathbb{Z}}$ is orthogonal if:

$$\langle\chi_d(t), \chi_s(t)\rangle = 0 \quad d\neq s \quad d,s\in\mathbb{Z} \tag{2.24}$$

where $\langle\chi_d(t), \chi_s(t)\rangle$ is the inner product of $\chi_d(t)$ and $\chi_s(t)$ that is given by:

$$\langle\chi_d(t), \chi_s(t)\rangle = \int_{-\infty}^{\infty} \chi_d(t)\chi_s^*(t)\,dt \tag{2.25}$$

If a set of functions satisfies the conditions to be a basis for a Hilbert space $H$ and satisfies the orthogonality condition, then it is an orthogonal basis for the Hilbert space $H$.

A scaling function $\phi(t)$ that satisfies the Riesz basis condition is capable of generating sets of basis functions in $H$. According to the Mallat theory, any set of basis functions in $H$ can be characterized by a pair of the form $(\Gamma, A_M)$, where:

- $A_M$ is an expanding matrix with its all eigenvalues $|\lambda_i| < 1$.
- $\Gamma$ is an invariant lattice of $A_M$ such that $A_M(\Gamma) \subseteq \Gamma$.

The matrix $A_M$ is called dilation matrix for $\Gamma$, and it has $|\det(A_M)| \in \mathbb{R}$. The dilation equation for any scaling function $\phi(t)$ can be written in terms of the pair $(\Gamma, A_M)$ as:

$$\phi(t) = \sum_{\gamma \in \Gamma} a_\gamma |\det(A_M)|^{1/2} \phi(t - \gamma) \tag{2.26}$$

The set $\{\phi(t - \gamma) : \gamma \in \Gamma\}$ is an orthogonal basis and the set of coefficients $\{a_\gamma\} \in \mathbb{R}$ or $\mathbb{C}$. The set $\{\phi(t - \gamma)\}$ can span a space $V(\phi)$. Also, high order spaces $V_j$ can be spanned by related orthogonal sets of the form $\{\phi(A_M^j t - \gamma) : \gamma \in \Gamma\}$, $j = 0, 1, 2 \ldots$ These sets of orthogonal functions are bases in $H$, which imply that $\{V_j\}_{j=0,1,2..} \in H$. Among all orthogonal basis functions in $H$, orthonormal basis functions play a very important role in constructing wavelet-based MRAs.

A set of orthogonal basis functions can be changed into a set of orthonormal basis functions using the Gram-Schmidt orthogonalization procedure. Using this procedure, a set of orthonormal basis functions can be constructed from another set of orthogonal basis functions. Let $\{\mu_1, \mu_2, \ldots, \mu_g\}$ be a set of orthogonal basis functions that span a linear space $\Lambda$. A set of orthonormal basis functions $\{\lambda_1, \lambda_2, \ldots, \lambda_g\}$ can be constructed with the same span as $\{\mu_i\}$ as:

$$\lambda_1 = \frac{\mu_1}{\|\mu_1\|} \tag{2.27}$$

Then, recursively evaluate:

$$\lambda_i = \frac{\mu_i - \nu_i}{\|\mu_i - \nu_i\|} \tag{2.28}$$

where $\nu_i$ is given by:

$$\nu_i = \sum_{p=0}^{g-1} \langle \lambda_p, \mu_i \rangle \lambda_p \tag{2.29}$$

The set of orthonormal basis functions $\{\lambda_1, \lambda_2, \ldots, \lambda_g\}$ obtained using the Gram-Schmidt procedure is capable of spanning the same linear space $\Lambda$ that is spanned by the orthogonal set of basis functions.

Note that if a set of orthonormal basis functions is generated from a Riesz basis, then the Riesz bounds become $A = B = 1$. These values of Riesz bounds satisfy the stability conditions, which have the following frequency-domain form:

$$\sum_{n=-\infty}^{\infty} \left| \hat{\phi}(\omega + 2\pi n) \right|^2 = 1 \tag{2.30}$$

where $\hat{\phi}(\omega)$ is the Fourier transform (FT) of $\phi(t)$.

A set of spaces $\{V_j\}_{j=0,1,2...}$ can be spanned by a set of orthonormal basis functions generated by a scaling function $\phi(t)$ such that:

$$\{\lambda_1, \lambda_2, \ldots, \lambda_g\} = \{\phi(A_M^j t - \gamma) : \gamma \in \Gamma\}, \quad j = 0, 1, 2 \ldots \tag{2.31}$$

Such a set of spaces has the following important characteristic:

$$\bigcap_{j=0,1,2...} V_j = \{0\} \quad \text{and} \quad \bigcup_{j=0,1,2...} V_j = L^2(\mathbb{R}) \tag{2.32}$$

This characteristic ensures the existence of $|\det(A_M)| - 1$ sets of basis functions that span a set of spaces $\{W_j\}_{j=1,2,...}$ as an orthogonal complement of $\{V_j\}_{j=1,2,...}$. This set of spaces ($\{W_j\}$) is spanned by sets of basis functions that are known as wavelet basis functions. These sets of basis functions are generated as an orthogonal complement to the set $\{\phi_j^k\}$ of basis functions as:

$$\psi_j(t) = \sum_{\gamma \in \Gamma} b_\gamma |\det(A_M)|^{1/2} \phi\left(A_M^j t - \gamma\right) \tag{2.33}$$

The function $\psi_j(t)$ is known as the wavelet function associated with $\phi_j(t)$ and the set of coefficients $\{b_\gamma\} \in \mathbb{R}$ or $\mathbb{C}$. The two sets of coefficients $\{a_\gamma\}$ and $\{b_\gamma\}$ are related as:

$$b_f = (-1)^f a_{\Gamma - f} \quad 0 < f \leq \Gamma \tag{2.34}$$

The basis functions $\{\psi(t)^k\}$ span a linear space $W$ that is orthogonal to $V$ as:

$$W = \text{clos}\left(\text{span}\left(\{\psi^k(t)\}\right)\right) \tag{2.35}$$

where span $(\{\psi_k(t)\})$ is given by:

$$\text{span}\left(\{\psi_k(t)\}\right) = \left\{\sum_{k \in \mathbb{Z}} \beta_k \psi_k(t) | \beta_k \in \mathbb{R} \text{ or } \mathbb{C}\right\} \tag{2.36}$$

The collection of these spanned spaces constructs an MRA. There are several common scaling functions that are capable of generating such sets of orthogonal and orthonormal basis functions, among them are the Daubechies and Haar scaling functions.

## 2.2.2  Semi-Orthogonal Wavelet Basis Functions

The semi-orthogonal wavelet basis functions are very close to orthogonal ones in spanning multiresolution spaces ($\{W_j\}$ and $\{V_j\}$). These basis functions are characterized using B-splines that are related to fractional differential operators. The semi-orthogonality condition forces wavelet spaces $\{W_j\}$ to be orthogonal to one another, which ensures that scaling spaces $\{V_j\}$ have the same orthogonal structure. However, scaling functions are selected to be generalized fractional B-splines, which are intimately related to a broad class of differential operators $\partial_\tau^\gamma$ with the $\gamma^{th}$ order derivative having a shift $\tau$.

The generalized fractional B-spline of degree $\alpha \geq 0$ ($\alpha \in \mathbb{R}$) and a shift of $\tau$ can be best defined as:

$$\hat{\beta}_\tau^\alpha(\omega) = \left(\frac{1 - e^{j\omega}}{-j\omega}\right)^{\frac{\alpha+1}{2} - \tau} \left(\frac{1 - e^{-j\omega}}{j\omega}\right)^{\frac{\alpha+1}{2} + \tau} \tag{2.37}$$

A set of scaling spaces $\{(V_j)_{SO}\}$ can be defined using these basis functions as:

$$(V_j)_{SO} = \text{clos}\left\{\sum_k \varepsilon_k \beta_\tau^\alpha (2^{-j}t - k) \middle| \varepsilon_k \in l^2(\mathbb{Z})\right\} \qquad (2.38)$$

The fractional $B$-spline basis functions have Riesz bounds such that:

$$0 < A_{SO} \leq \sum_{k=-\infty}^{\infty} \left|\hat{\beta}_\tau^\alpha(\omega + 2\pi k)\right|^2 \leq B_{SO} < \infty \qquad (2.39)$$

where $\hat{\beta}_\tau^\alpha(\omega)$ is the FT of $\beta_\tau^\alpha(t)$. Complementary basis functions can be defined using the fractional $B$-spline scaling function $\beta_\tau^\alpha(t)$ as:

$$\psi_\tau^\alpha(t) = \sum_{k \in \mathbb{Z}} w_k \beta_\tau^\alpha (2t - k) \quad w_k \in l^2(\mathbb{Z}) \qquad (2.40)$$

In the frequency domain, this relation takes the following form:

$$\hat{\psi}_\tau^\alpha(\omega) = W(\omega)\frac{\hat{\beta}_\tau^\alpha(\omega)}{2} \qquad (2.41)$$

A wavelet spanned space $W_{SO}$ can be defined as:

$$(W_j)_{SO} = \text{span}\left\{\psi_\tau^\alpha(2^{-j}t)\right\} \qquad (2.42)$$

The two spaces $(V_j)_{SO}$ and $(W_j)_{SO}$ are spanned by nonorthonormalized $B$-splines basis functions. These features make semi-orthogonal sets of basis functions always behave asymptotically as a fractional differential operator. Consequently, analyzing a signal $f(t)$ with wavelet semi-orthogonal basis functions yields samples of the operator $\partial_\tau^\gamma$ applied to a smoothed version of $f(t)$:

$$\langle f(t), \psi(t-k)\rangle = \partial_\tau^\gamma \{\xi * f\}[k] \qquad (2.43)$$

where $\xi[k]$ is a smoothing function defined in the frequency domain as:

$$\hat{\xi}(\omega) = \frac{\hat{\psi}(\omega)}{\hat{\partial}_\tau^\gamma(-\omega)} \qquad (2.44)$$

where $\hat{\partial}_\tau^\gamma(\omega)$ is the FT of $\partial_\tau^\gamma$ operator, and is given by:

$$\hat{\partial}_\tau^\gamma(\omega) = (-j\omega)^{\frac{\gamma}{2} - \tau}(j\omega)^{\frac{\gamma}{2} + \tau} \qquad (2.45)$$

It is to be noted that spaces spanned by sets of semi-orthogonal basis functions are orthogonal $((W_j)_{SO} \perp (V_j)_{SO})$ for some values of $\gamma$ that satisfy the following condition:

$$\gamma \approx \alpha + 1 \qquad (2.46)$$

## 2.2.3  Bi-Orthogonal Wavelet Basis Functions

Orthogonal and semi-orthogonal basis functions provide series expansions of signals that have finite energy with a general form as:

$$f(t) = \sum_{j,k} d_{j,k} \psi_{j,k}(t) \quad f(t) \in L^2, \quad d_{j,k} \in \mathbb{R} \text{ or } \mathbb{C} \tag{2.47}$$

The set of expansion coefficients $\{d_{j,k}\}$ has information about the time-frequency structure of the expanded signal $f(t)$. These coefficients can be determined as:

$$d_{j,k} = \int_{-\infty}^{\infty} f(t) \psi_{j,k}(t) \, dt = (E_\psi f) \left( \frac{k}{2^j}, \frac{1}{2^j} \right) \tag{2.48}$$

where $(E_\psi f)(a, b)$ is the integral wavelet transform (*IWT*) given as:

$$(E_\psi f)(a, b) = \frac{1}{\sqrt{a}} \int_{-\infty}^{\infty} f(t) \psi \left( \frac{t-b}{a} \right) dt, \quad \text{with } (a, b) = \left( \frac{k}{2^j}, \frac{1}{2^j} \right) \tag{2.49}$$

In such expansion the function $\psi_{j,k}(t)$ is used to analyze the signal $f(t)$ as well as to provide time-locations of different frequencies present in it. Moreover, the spanned spaces $(W_j)_\psi$ and $(V_j)_\phi$ have to satisfy the orthogonality condition for every scale $j$ such that:

$$\left( W_j \right)_\psi \perp \left( V_j \right)_\phi \tag{2.50}$$

The fundamental concept of bi-orthogonal wavelet basis functions is based on selecting two scaling functions ($\phi(t)$ and $\tilde{\phi}(t)$) that are dual to each other such that:

$$\left\langle \phi(t-k), \tilde{\phi}(t-m) \right\rangle = \delta_{k,m} \quad k, m \in \mathbb{Z} \tag{2.51}$$

If wavelet functions ($\psi(t)$ and $\tilde{\psi}(t)$) are associated with $\phi(t)$ and $\tilde{\phi}(t)$, respectively, then these wavelet functions are dual such that:

$$\left\langle \psi(t-l), \tilde{\psi}(t-q) \right\rangle = \delta_{l,q} \quad l, q \in \mathbb{Z} \tag{2.52}$$

These dual functions can span four different spaces at each scale, which are $(V_j)_\phi$, $(\tilde{V}_j)_\phi$, $(W_j)_\psi$, and $(\tilde{W}_j)_\psi$. Some of these spaces satisfy the orthogonality conditions such that:

$$V_j \cap W_j = \{0\} \quad \text{Nonorthogonal} \tag{2.53}$$

$$\tilde{V}_j \cap \tilde{W}_j = \{0\} \quad \text{Nonorthogonal} \tag{2.54}$$

$$\tilde{V}_j \perp W_j \quad \text{Orthogonal} \tag{2.55}$$

$$\tilde{W}_j \perp V_j \quad \text{Orthogonal} \tag{2.56}$$

Bi-orthogonal scaling and wavelet functions can generate different sets of basis functions for analyzing and synthesizing signals. These different sets of basis functions can offer better representation than orthogonal and semi-orthogonal basis functions for certain types of signals.

## 2.2.4  **Shift-Orthogonal Wavelet Basis Functions**

Shift-orthogonal wavelet basis functions are sets of basis functions that span spaces, which are orthogonal with respect to translations in each scale. However, these spaces are not orthogonal with respect to dilations across scales. These basis functions are generated by scaling and wavelet functions that satisfy the duality principle (bi-orthogonality). The construction of shift-orthogonal scaling functions is based on selecting any two analysis and synthesis scaling functions $\zeta_a(t)$ and $\zeta_s(t)$ and defining an autocorrelation sequence $ac[k]$ as:

$$ac_{s,a}[k] = \langle \zeta_a(t-k), \zeta_s(t) \rangle = \left( \zeta_a^T * \zeta_s \right)[k] \tag{2.57}$$

A synthesis scaling function $\phi_s(t)$ can be constructed as an orthogonalized version of $\zeta_s(t)$ as:

$$\phi_s(t) = \sum_{k \in \mathbb{Z}} (ac_{s,s})^{1/2}[k] \zeta_s(t-k) \tag{2.58}$$

where $(ac_{s,s})^{1/2}[k]$ is given by:

$$(ac_{s,s})^{1/2}[k] \xleftarrow{\;DTFT\;} \frac{1}{\sqrt{\widehat{ac}_{s,s}(\omega)}} \tag{2.59}$$

The analysis scaling function $\phi_a(t)$ can be constructed as the dual of $\phi_s(t)$ as:

$$\phi_a(t) = \tilde{\phi}_s(t) = \sum_{k \in \mathbb{Z}} \left( (ac_{s,s})^{1/2}[k] * \left( ac_{a,s}^T \right)^{-1}[k] \right) \zeta_a(t-k) \tag{2.60}$$

where $\left( ac_{a,s}^T \right)^{-1}[k]$ is given by:

$$\left( ac_{a,s}^T \right)^{-1}[k] \xleftarrow{\;DTFT\;} \frac{1}{\widehat{ac}_{a,s}(\omega)} \tag{2.61}$$

The definition of the analysis scaling function $\phi_a(t)$ indicates that it is not orthogonal to its own shifts. This feature can be expressed as:

$$\langle \phi_a(t), \phi_a(t-k) \rangle = ac_\phi[k] \neq 0 \tag{2.62}$$

The sequence $ac_\phi[k]$ is an autocorrelation sequence that can be determined as:

$$ac_\phi[k] = \left( ac_{s,s} * \left( ac_{a,s} * ac_{a,s}^T \right)^{-1} * ac_{a,a} \right)[k] \tag{2.63}$$

Two wavelet functions can be constructed using both $\zeta_a(t)$ and $\zeta_s(t)$ as:

$$\psi_s(t) = \sum_{k \in \mathbb{Z}} p[k] \zeta_s(2t-k) \tag{2.64}$$

$$\psi_a(t) = \sum_{k \in \mathbb{Z}} \hat{p}[k] \zeta_a(2t-k) \tag{2.65}$$

where the two sequences $p[k]$ and $\hat{p}[k]$ are extended dual filter sequences. The derivation of both $p[k]$ and $\hat{p}[k]$ is detailed in reference. The two wavelet functions

$\psi_a(t)$ and $\psi_s(t)$ satisfy the duality condition. Also, $\psi_a(t)$ wavelet function is not orthogonal to its own shifts, where a similar autocorrelation sequence $ac_\psi[k]$ can be derived as a nonzero sequence. The spanned spaces by shift-orthogonal basis functions are not orthogonal to each other such that:

$$V_j \cap W_j = \{0\} \quad \text{Nonorthogonal} \qquad (2.66)$$

$$\tilde{V}_j \cap \tilde{W}_j = \{0\} \quad \text{Nonorthogonal} \qquad (2.67)$$

$$\tilde{V}_j \cap W_j = \{0\} \quad \text{Nonorthogonal} \qquad (2.68)$$

$$\tilde{W}_j \cap V_j = \{0\} \quad \text{Nonorthogonal} \qquad (2.69)$$

Different types of wavelet basis functions have been constructed for various applications in signal and image processing areas. The diversity of approaches employed in constructing such basis functions has been motivated to meet the requirements of various applications. In general, wavelet basis functions are capable of spanning spaces through integer dilations and translations. Furthermore, the collection of the spaces spanned by one set of basis functions defines its associated MRA. Each defined MRA can support only a uniform sampling-reconstruction process due to the dyadic ($2^n$ dilations) nature of these spanned spaces. The review provided in the previous section has briefly described different types of available wavelet basis functions and the main characteristics of their associated MRAs.

## 2.3 SAMPLING PROCESS AS A MULTIRESOLUTION ANALYSIS (MRA)

There has been extensive research devoted to the sampling theorem, with the most important piece of work being the classical sampling theorem. The classical sampling theorem, also known as Whittaker-Kotelńikov-Shannon (WKS) theorem, states that a low pass signal continuous-time (CT) band-limited to a frequency band of $(-f_0, f_0)$ can be reconstructed perfectly from its samples taken uniformly at no less than the Nyquist rate of $2f_0$ samples/sec. Another alternative condition for successful sampling and reconstruction of CT band-limited signals is introduced in terms of constraining the lower bound on the sampling density for perfect reconstruction. According to the classical WKS sampling theorem, a CT signal $x_c(t)$ band-limited to $(-\Omega_0, \Omega_0)$, such that $X_c(\Omega) = 0$, $|\Omega| > \Omega_0$ ($\Omega_0 = 2\pi f_0$), can be perfectly recovered from samples spaced by $T_s$ ($T_s \leq \pi/\Omega_0$). It is to be noted that $X_c(\Omega)$ is the continuous time Fourier transform (CTFT) of $x_c(t)$ as:

$$X_c(\Omega) = \int_{-\infty}^{\infty} x_c(t) e^{-j\Omega t} dt \qquad (2.70)$$

The ideal sampled version of $x_c(t)$ is a sequence of discrete points located at multiples of the sampling period $T_s$, and can be defined as:

$$x[n] = \sum_n x_c(t) \delta(t - nT_s) \qquad (2.71)$$

Ideal reconstruction of the CT signal from its samples ($x[n]$) can be done as:

$$x_c(t) = \sum_n x[n] \operatorname{sinc}(t - nT_s) \tag{2.72}$$

where $\operatorname{sinc}(t) = \sin(\pi t)/\pi t$. This form of sampling is known as the uniform (periodic) sampling. It should be noted that there are other forms of sampling that employ a non-uniform sampling approach. Such sampling forms have been employed in many areas of signal and image processing applications.

An important extension to the sampling theorem was contributed by Kramer, who proposed the use of generalized integral transforms rather than using the Fourier transform only. Also, Kramer's extension has led to an important interpretation of the sampling process that can be stated as follows: sampling a CT signal $x_c(t)$, where $x_c(t) \in C_b[0, T_s]$, is equivalent to extracting a set of $N$ real-valued parameters from that signal $x_c(t)$. The space $C_b[0, T_s]$ is a space of continuous and bounded functions on $[0, T_s]$. The span of a collection of $N$ linearly independent basis vectors $\{v_1, v_2..., v_N\}$ ($v_k \in C_b[0, T_s]$) will create an $N$-dimensional subspace of $C_b[0, T_s]$. Furthermore, if the real-valued parameters are considered weights to these basis vectors, any vector in this subspace can be expressed as:

$$\hat{x} = \sum_{k=1}^{N} c_k v_k, \quad c_k = \mathbb{R} \tag{2.73}$$

where the set $\{c_k\}$ represents the projection of $\hat{x}$ on individual subspaces spanned by the basis vectors. The vector $\hat{x}$ can be considered a reconstruction of the signal $x_c(t)$. Such an interpretation is basically an approximation of $x_c(t)$ in the $N$-dimensional subspace $V$ of $C_b[0, T_s]$. Moreover, highly accurate reconstruction can be achieved if the signal $x_c(t)$ lies in the subspace $V$.

The classical sampling theorem can be interpreted in terms of an orthogonal projection of the sampled signal into a function subspace $V$. A set of band-limited continuous-time signals $\{CT(t)\}$ can be expressed as:

$$\{CT(t)\} \subseteq B_{2\Omega_o}^2, \quad \Omega_o > 0 \tag{2.74}$$

It is to be noted that the function subspace $V$ is an $L^2$-closure that is:

$$C_{Sa,\{CT(t)\}} = \operatorname{clos}_{L^2}\langle Sa(t - ka) : k \in \mathbb{Z} \rangle \tag{2.75}$$

This can help in defining a mapping for the function space $L^2$ into the subspace $V_{Sa,\{CT(t)\}}$. Such mapping can be defined using a sampling operator $(O_a x_c)(t)$ as:

$$(O_a x_c)(t) = \sum_{k=-\infty}^{\infty} x_c(ka) Sa_{s;a}(t - ka) \tag{2.76}$$

where $Sa_{s;a}(t)$ is the Shannon sampling function given by:

$$Sa_{s;a}(t) = \frac{\sin\left(\pi \dfrac{t}{a}\right)}{\pi \dfrac{t}{a}} \tag{2.77}$$

This mapping provides a fairly accurate signal representation for $x_c(t) \in \{CT(t)\}$ due to the conditions imposed on the spacing $a$ between samples. In order to obtain a more accurate representation of the signal $x_c(t)$, an orthogonal projection mapping $(L_a x_c)(t)$ can be used. This mapping is defined as:

$$(L_a x_c)(t) = \sum_{k=-\infty}^{\infty} c_k(ka) Sa_{s;a}(t - ka) \tag{2.78}$$

such that:

$$\int_{-\infty}^{\infty} (x_c(t) - (L_a x_c)(t)) Sa_{s;a}(t - ka) \, dt = 0 \tag{2.79}$$

The coefficients $\{c_k\}$ are defined as:

$$c_k = \int_{-\infty}^{\infty} x_c(t) Sa_{s;a}(t - ka) \, dt, \quad k \in \mathbb{Z} \tag{2.80}$$

The relation between the two mappings can be investigated under the following conditions:

- The signal $x_c(t) \in \{CT(t)\}$.
- The sampling function $Sa(t)$ is capable of generating an orthonormal family on the samples set $\{ka\}$, $k \in \mathbb{Z}$.
- The spacing between samples $a$ has to satisfy $0 < a \le \dfrac{\pi}{\Omega_o}$.
- The Fourier transform of the sampling function $Sa(t)$ is such that:

$$\hat{Sa}(\Omega) = \int_{-\infty}^{\infty} Sa(t) e^{-j\Omega t} \, dt = 1 \; for \; |\Omega| \le \Omega_o \tag{2.81}$$

The orthogonal projection mapping becomes:

$$(L_a x_c)(t) = \sum_{k=-\infty}^{\infty} \left( \int_{-\infty}^{\infty} x_c(t) Sa_{s;a}(t - ka) \, dt \right) Sa_{s;a}(t - ka) \tag{2.82}$$

Using the Parseval's identity, equation (2.40) can be written as:

$$(L_a x_c)(t) = \sum_{k=-\infty}^{\infty} \left( \frac{1}{2\pi} \int_{-\Omega_o}^{\Omega_o} X_c(\Omega) \hat{Sa}^*(\Omega) e^{jka\Omega} \, d\Omega \right) Sa_{s;a}(t - ka) \tag{2.83}$$

$$(L_a x_c)(t) = \sum_{k=-\infty}^{\infty} \left( \frac{1}{2\pi} \int_{-\Omega_o}^{\Omega_o} X_c(\Omega) e^{jka\Omega} \, d\Omega \right) Sa_{s;a}(t - ka) \tag{2.84}$$

The term $\dfrac{1}{2\pi} \int_{-\Omega_o}^{\Omega_o} X_c(\Omega) e^{jka\Omega} \, d\Omega$ is an inverse CTFT that is equal to $x_c(ka)$ such as:

$$\frac{1}{2\pi} \int_{-\Omega_o}^{\Omega_o} X_c(\Omega) e^{jka\Omega} \, d\Omega = x_c(ka) \tag{2.85}$$

The equation of the orthogonal mapping becomes:

$$(L_a x_c)(t) = \sum_{k=-\infty}^{\infty} x_c(ka) Sa_{s;a}(t - ka) = (O_a x_c)(t) \tag{2.86}$$

Equation (2.86) shows that the two mappings perform the same decomposition under certain conditions. Among these conditions is the capability of the sampling function to generate an orthonormal family over the sampling set. This condition indicates that the sampling function has to satisfy specific conditions required by a scaling function for generating a set of orthonormal basis functions.

The sampling mapping and the orthogonal projection mapping are identical if the sampling function $Sa(t)$ can generate orthonormal basis functions over the samples set $\{ka\}$, $k \in \mathbb{Z}$. This condition can be verified using the concept of multi-dimensional approximation. Starting with the classical sampling theorem, a signal $x_c(t)$ band-limited to $(-\Omega_o, \Omega_o)$ ($x_c(t) \in CT(t)$) can be reconstructed or approximated from its samples as:

$$x_c(t) = \sum_{n=-\infty}^{\infty} x(nT_s) \frac{\sin \Omega_o(t - nT_s)}{\Omega_o(t - nT_s)} \tag{2.87}$$

where $T_s = \pi/\Omega_o = a$ and $t \in \mathbb{R}$. If $\Omega_o$ is allowed to vary as $\Omega_o = 2^m \pi$, $m \in \mathbb{Z}$, then this can be viewed as a setting of an MRA. The sinc function can be defined then as a scaling function of the MRA such that it can generate a collection of linear independent basis vectors $\{v_1, v_2 ..., v_N\}$. The scaling function satisfies the following dilation equation:

$$\phi(t) = \sum_k \frac{\sin \pi k/2}{\pi k/2} \phi(2t - k) \tag{2.88}$$

This setting encompasses the classical sampling theorem in an MRA context. Also, it creates a connection between the WSK classical sampling theorem and the wavelet theory. The relation between wavelet theory and MRA has led to an important conclusion that there exists a sampling function $\phi(t)$, which can provide a sampling expansion of any CT signal $x_c(t) \in V_0$. The conditions required for a function to generate orthonormal bases for an MRA are described as follows:

- The function has to be a real and a continuous function with a decaying property such that:

$$\lim_{|t| \to \infty} \phi(t) = 0, \quad t \in \mathbb{R} \tag{2.89}$$

- The integer translations of $\phi(t)$ of the form $\{\phi(t - l)\}$, $l \in \mathbb{Z}$ form an orthonormal basis for a subspace $V_0$ of $L^2(\mathbb{R})$.
- The MRA generated by $\phi(t)$ of closed subsets $\{V_m^2\}_{m \in \mathbb{Z}}$ of $L^2(\mathbb{R})$ has to satisfy:

$$0 ... \subset V_{-1} \subset V_0 \subset V_1 \subset V_2 ... \subset V_m \subset L^2(\mathbb{R}) \tag{2.90}$$

$$x_c(t) \in V_m \Leftrightarrow x(2t) \in V_{m+1} \tag{2.91}$$

$$\bigcap_{m \in \mathbb{Z}} V_m = 0, \quad \overline{\bigcup_{m \in \mathbb{Z}} V_m} = L^2(\mathbb{R}) \tag{2.92}$$

- Since $\phi(t) \in V_1$, there exists a sequence $\{a_k\}$ of length $N$ such that:

$$\phi(t) = \sum_{k=0}^{N-1} a_k \phi(2t - k) \tag{2.93}$$

## 2.4 SAMPLING FORMS

Discrete-time signals can arise in different ways, but the most common way is the representation of continuous-time signals. It is remarkable that under reasonable conditions and constraints, a continuous-time signal is accurately represented by its values at discrete points in time. These points in time are selected so that the sampled continuous-time signal can be recovered perfectly without any distortion or loss of information. Two forms for selecting time instants are commonly used in signal processing applications, uniform sampling, and nonuniform sampling. In the uniform sampling, time instants are selected with equidistant spacings. On the other hand, nonuniform sampling involves selecting time instants with variable spacings. Figure 2.1 shows a signal $x_c(t)$ sampled both uniformly and nonuniformly.

The following subsections provide more insight about these sampling forms.

### 2.4.1 Uniform Sampling

Uniform sampling, also known as periodic sampling, is the most common form of sampling and is used in a wide range of signal and image processing applications.

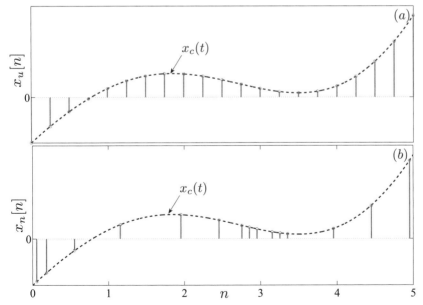

**Figure 2.1** Sampling a continuous signal (CT) $x_c(t)$: (a) the uniform sampling of $x_c(t)$ to produce the discrete signal $x_u[n]$ and (b) the nonuniform sampling of $x_c(t)$ to produce the discrete signal $x_n[n]$.

Moreover, all mathematical approaches of the sampling theorem are initially developed considering the uniform sampling. In this sampling form, samples are taken at a constant rate resulting in a constant spacing between successive samples.

Let $x_d[n]$ be the sampled form of a band-limited signal $x_c(t)$, where $x_c(t) \in \mathcal{B}^2_{2\Omega_o}(\mathbb{R})$. The space $\mathcal{B}^2_{2\Omega_o}(\mathbb{R})$ is the space of all signals band-limited to $\Omega_o$, and is defined as:

$$\mathcal{B}^2_{2\Omega_o}(\mathbb{R}) = \{x_c(t) \in L^2(\mathbb{R}) : \sup(X_c(\Omega)) \subseteq [-\Omega_o, \Omega_o]\} \tag{2.94}$$

A discrete signal $x_d[n]$ is defined by samples taken with an equidistant spacing of $T_s$ such that:

$$\forall n, n \in \mathbb{Z}, x_d[n] = x_c(nT_s) \tag{2.95}$$

The equidistant samples can be created by using a train of impulses located at integer multiples of $T_s$. Such a train of impulses can be defined as:

$$p(t) = \sum_{n=-\infty}^{\infty} \delta(t - nT_s) \tag{2.96}$$

where $\delta(t)$ is the Dirac delta function, defined by the following properties:

- Unit area:

$$\int_{-\infty}^{\infty} \delta(t) dt = 1 \tag{2.97}$$

- Sifting property: for a function $f(t)$ continuous at $t = \tau$

$$\int_{-\infty}^{\infty} f(t)\delta(t - \tau) dt = f(\tau) \tag{2.98}$$

Let $x_s(t)$ be defined as follows: for $x_c(t)$ continuous at all $t = nT_s$

$$x_s(t) = x_c(t).p(t) = x_c(t) \sum_{n=-\infty}^{\infty} \delta(t - nT_s) \tag{2.99}$$

using the sifting property of $\delta(t)$, $x_s(t)$ becomes:

$$x_s(t) = \sum_{n=-\infty}^{\infty} x_c(nT_s)\delta(t - nT_s) \tag{2.100}$$

The discrete signal $x_d[n]$ can be defined as:

$$x_d[n] = \int_{nT_s-\varepsilon}^{nT_s+\varepsilon} x_s(t) dt, \quad \varepsilon \in [0, T_s] \tag{2.101}$$

The spacing between adjacent samples has to be less than a value known as the Nyquist rate $T_Q$ in order to avoid aliasing and to facilitate accurate reconstruction of the CT signal from its samples. The Nyquist rate is related to the highest frequency component present in the sampled CT signal such that:

$$T_Q \leq \frac{1}{2f_{hst}} \Rightarrow f_Q \geq 2f_{hst} \tag{2.102}$$

where $f_{hst}$ is the highest frequency component present in the sampled CT signal $x_c(t)$ in cycles/sec. and $f_Q$ is the Nyquist frequency. If the Nyquist condition is met, then the CT signal $x_c(t)$ can be recovered perfectly from its samples. This perfect recovery is represented using a low-pass filter with a cut-off frequency of $\Omega_{CF} = \dfrac{\Omega_s}{2}$.

## 2.4.2  Nonuniform Sampling

A uniform sampling case turns into a nonuniform sampling one if $T_s$ fails to remain constant. This condition implies that the spacing between adjacent samples will no longer be constant. This nonuniform sampling form can be found in some engineering and geophysics data acquisition applications. Although nonuniform sampling is the norm rather than the exception, it has not received the same attention as the uniform sampling in signal and image processing applications. One of the main reasons for the lack of attention to the nonuniform sampling is the difficulty encountered when performing the Fourier analysis. For example, sequences with nonequidistant samples affect the convergence of the infinite series required for applying Fourier analysis. From a theoretical point of view, the nonuniform sampling problem has inspired many to develop few elegant, but mostly not constructive, theorems. However, there exist some algorithms for approximating or reconstructing one-dimensional and two-dimensional band-limited signals from their nonuniform samples. These algorithms include polynomial and spline methods.

The most popular work for considering the nonuniform sampling is the theory developed by Levinson that is based on *Lagrange interpolation functions* such that: Let $h(t)$ be an entire function with zeros only at $\{t_n : n \in \mathbb{Z}\}$ defined as:

$$h(t) = (t - t_o) \prod_n \left(1 - \frac{t}{t_n}\right)\left(1 - \frac{t}{t_{-n}}\right) \tag{2.103}$$

where $t_o$ is an arbitrary reference point. Using the defined Lagrange interpolation function $h(t)$, a band-limited signal can be expanded using these functions. In other words, a band-limited signal can be recovered from its nonuniform samples using this type of interpolation functions such that:

$$x_c(t) = \sum_{n \in \mathbb{Z}} x_c(t_n) h(t_n) \tag{2.104}$$

A popular pertinent case of nonuniform sampling is the case when a set of nonuniform samples can be divided into subsets with a condition that the number of these subsets is finite such that:

$$|t_n - t_m|_k > \alpha_k, \quad m \neq n, \quad m, n \in \mathbb{Z}, \quad k < \infty \tag{2.105}$$

where $\{\alpha_k\}_{k \in \mathbb{Z}} > 0$. This condition is known as the relative separation principle, which defines the condition over the spacing between samples in each subset $k$. If there are sets of separable nonuniform samples, then there exists a lower uniform Beurling density condition. Moreover, if such sets have a repetitive nature (periodicity), then this type of nonuniform sampling becomes the nonuniform recurrent sampling.

## 2.4.3 Nonuniform Recurrent Sampling

In some practical applications, aliasing can cause problems in recovering periodic signals from their uniform samples. One of the possible methods to avoid such problems is using nonuniform sampling, in particular nonuniform recurrent sampling. In this form of sampling, nonuniform samples are divided into groups (subsets) of $N$ samples each. These sample groups have a recurrent period of $T$, which can be related to $N$ by:

$$T = NT_Q \tag{2.106}$$

where $T_Q$ satisfies:

$$T_Q \leq \frac{T_m}{2} \tag{2.107}$$

When nonuniform samples have a minimum spacing that satisfies the Nyquist condition, recovery of the signal from such samples is possible. This has been described by the lower uniform Beurling density approach. Also, when nonuniform samples can be divided into a finite number of subsets, equation (2.24) can be employed for recovering signals from their nonuniform recurrent samples. The locations of nonuniform recurrent samples in the sample group (subset) $d$ can be stated as $\{t_p\}_{p=0,1,2,\cdots,N-1}$ relative to a start time for that group.

If the nonuniform recurrent samples of the signal $x_c(t)$ satisfy the lower uniform Beurling density condition, then it can be recovered from these samples provided the average sampling period is smaller than the Nyquist rate. The average sampling period for the case of nonuniform recurrent sampling can be defined as:

$$T_{avg} = \lim_{n \to \infty} \frac{t_n - t_{n-1}}{n} \tag{2.108}$$

Figure 2.2 shows a band-limited signal sampled in a nonuniform recurrent manner.

Recovering the signal $x_c(t)$ from its nonuniform samples located at $\{t_p\}$ can be done using Lagrange interpolation functions as:

$$x_c(t) = \sum_{n=-\infty}^{\infty} x_c(t_n) \frac{G(t)}{G'(t_n)(t - t_n)} \tag{2.109}$$

where

$$G(t) = (t - t_o) \prod_p \left(1 - \frac{t}{t_p}\right) \tag{2.110}$$

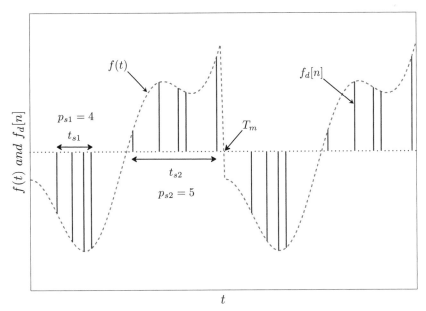

**Figure 2.2**    Sampling a periodic CT signal $f(t)$ in a nonuniform recurrent manner using two sample groups at each period.

and

$$G'(t_n) = \frac{dG(t)}{dt}\bigg|_{t=t_n} \tag{2.111}$$

The recovery of a signal from its samples (uniform or nonuniform) can be carried out by interpolation processes using filter banks. One of the most efficient structures of filter banks is the quadrature-mirror filter banks, which can define a MRA-synthesis structure. The previous discussion of the relation between the sampling process and the MRA provided an important relation between sampling a function and analyzing the same function. Moreover, the sampling process can be realized accurately using an MRA structure. Also, the discussion indicated that there exists a strong connection between sampling a function and processing the same function using the wavelet analysis. The next section provides further interpretation of sampling a signal using wavelet basis functions.

## 2.5  WAVELET SAMPLING THEORY

A continuous-time (CT) signal $x_c(t)$ can be recovered from its samples created by a scaling function. A scaling function $\phi(t)$ has to be capable of generating basis functions that span a set of closed spaces $\{V_m\}$, $m \in \mathbb{Z}$. Moreover, the density and completeness conditions require an orthogonal complement space $W_m$ for each space

$V_m$. Each orthogonal complement space $W_m$ is spanned by another set of basis functions generated by $\phi_m(t)$. Such a set of basis functions defines a wavelet function $\psi(t)$ associated with $\phi(t)$ as:

$$\psi(t) = \sum_{k=1}^{N} (-1)^k a_{N-k} \phi(2t - k) \tag{2.112}$$

Basis functions required to span each $W_j$ can be generated at each scale $j$ by integer translations of the wavelet function $\psi(t)$, and are known as wavelet basis functions as:

$$\{\psi_{j,k}\} = 2^{j/2} \psi(2^j t - k) \tag{2.113}$$

Each orthogonal complement space $W_m$ can be defined as a linear span of wavelet basis functions as:

$$W_j = \operatorname*{span}_{k} \{\psi_{j,k}\} \tag{2.114}$$

The same applies for the space $V_m$ that is a linear span of scaling basis functions as:

$$V_j = \operatorname*{span}_{k} \{\phi_{j,k}\} \tag{2.115}$$

A collection of scaling spaces $V_m(\phi)$ and wavelet spaces $W_M(\psi)$ constitutes an MRA. It is to be noted that wavelet and scaling spaces have to be dense and complete in $L^2(\mathbb{R})$, that is:

$$MRA = \left\{ V_m \overset{J-1}{\underset{m=0}{\oplus}} W_m \right\} \tag{2.116}$$

Density and completeness conditions of both spaces can be used to relate them with MRA as the scale $j$ changes. This can be translated in constructing scaling space $V_q(\phi)$ as:

$$V_q(\phi) = V_{q-1}(\phi) \oplus W_{q-1}(\psi) \tag{2.117}$$

One of the interpretations of sampling a signal is an $N$-dimensional approximation case. If a CT signal $x_c(t)$ is contained in a space $V_q$, then it can be expanded using basis functions that are generated at scale $q$ as:

$$x_c(t) = \sum_{k \in \mathbb{Z}} (c_\phi)_k \phi(t - k) + \sum_{j=0}^{q} \sum_{k \in \mathbb{Z}} (c_\psi)_{k,j} \psi_j(t - k) \tag{2.118}$$

where coefficient sets $\{(c_\phi)_k\}$ and $\{(c_\psi)_{k,j}\}$ represent projecting the signal on scaling and wavelet spaces, respectively. These coefficient sets can be determined using inner product operations as:

$$\{(c_\phi)_k\} = \langle x_c(t), \tilde{\phi}(t - k) \rangle \tag{2.119}$$

$$\{(c_\psi)_{k,j}\} = \langle x_c(t), \tilde{\psi}_j(t - k) \rangle \tag{2.120}$$

where $\tilde{\phi}(t)$ is the dual scaling function and $\tilde{\psi}_j(t)$ is the dual wavelet function.

The previous discussion indicates that the projection of a CT signal $x_c(t)$ can represent a sampling-reconstruction process. This can be generalized to an approximation case in terms of a wavelet-based MRA as:

$$x_c(t) = \sum_{j=0}^{q-1}\sum_{k\in\mathbb{Z}}\langle x_c(t), \tilde{\phi}_j(t-k)\rangle \phi_j(t-k) \tag{2.121}$$

The term $\langle x_c(t), \tilde{\phi}_j(t-k)\rangle$ represents a generalized sampling of the CT signal $x_c(t)$, where the set of basis functions $\{\phi_j(t-k)\}_{k\in\mathbb{Z}}$ span an approximation space at each scale $j$.

## 2.6  SUMMARY

There are different types of wavelet basis functions that are capable of spanning spaces to construct MRAs. The majority of known MRAs are based on dyadic structure that can only support uniform sampling. Wavelet basis functions can be designed using the dilation matrix-based method, convolution of dual basis functions, fractional *B-spline* basis functions and, the autocorrelation sequence-based method. These methods of designing wavelet basis functions guarantee different combinations of scaling and wavelet spaces within dyadic MRA structures. The need for nondyadic MRAs in applications that include modeling inverters has motivated the development of a new type of scaling and wavelet basis functions to construct such types of MRA.

For the last 50 years, the sampling theorem has been a subject for extensive research due to its widespread applications. Mathematical derivations for the sampling process using functional space analysis and the approximation theory have been developed. Due to new applications, different forms of sampling emerged and derivations had to be generalized to accommodate these new sampling forms. One of these is the nonuniform recurrent sampling form that is applied in periodic signal sampling and reconstruction. Furthermore, the interest of applying wavelet MRA in signal processing areas has led to the wavelet sampling theory. Several important contributions have been achieved in this aspect, in particular the representation of sampling-reconstruction process as a wavelet-based MRA-synthesis.

The concepts developed for the sampling theorem can be employed to build mathematical models for various system functions. Among such functions are the switching functions, in particular, for power electronic inverters. Power electronic switch-mode inverters carry out switching actions for different time intervals. The main objective of such switching actions is to synthesize a sinusoidal signal on a high power level. Different aspects of operating these systems are still not validated using the conventional interpretation of the inverter operation. There is a need for a mathematical modeling of power electronic switch-mode inverter functions that are based on concepts of the sampling theorem. Such a model can be used as a foundation to develop a new modulation technique that is capable of operating inverters for improved output performances.

# Chapter 3

# Modeling of Power Inverters

## 3.1 INTRODUCTION

Inverters are mostly operated using multiswitching techniques to eliminate or reduce as much energy distributed in the output harmonic components as possible. Multiswitching techniques can be realized through a modulation process that determines time instants at which switching element(s) change their status (ON to OFF or OFF to ON). Various approaches with different schemes of implementation have been developed and tested to carry out multiswitching techniques for improving inverter performance. Although extensive research has focused on operating and controlling inverters, there remains a scope to develop accurate mathematical models capable of representing the inverter function. Moreover, existing modulation techniques are optimized to meet load requirements with limited correlation with any existing inverter model. This approach of operating inverters has caused several limitations of modulated inverters performance. These limitations include the spectral distribution of inverter output harmonic components, the jitter phenomenon when using rectangular pulse carrier signals, and impacts on the output due to changing the switching strategy.

This chapter presents an approach for modeling single-phase ($1\phi$) and three-phase ($3\phi$) inverters using the nonuniform sampling-reconstruction concepts. This modeling approach aims to develop new mathematical models for $1\phi$ and $3\phi$ inverters with the following capabilities:

- Providing a mathematical tool for verifying the impact on inverter outputs due to applying a switching strategy.
- Providing a basis for novel modulation techniques that can be correlated with the inverter model.
- Providing new approaches for realizing new control techniques for inverter outputs.

In general, the common assumption in modeling power electronic converters has been based on time-averaging the switching actions over one cycle of a reference-

*An Introduction to Wavelet Modulated Inverters*, by S.A. Saleh and M. Azizur Rahman
Copyright © 2011 Institute of Electrical and Electronics Engineers

modulating signal (usually a sinusoidal signal). This assumption has been realized in three main models, which include steady-state models derived using circuit theory, operational models derived from numerical data, and models derived from a unified converter theory.

The earliest steady-state circuit modeling approach of switching converters is based on the small-signal analysis concept. In this model, any switching circuit is assumed to be piece-wise linear and its response is determined for any small perturbation of steady-state operating conditions. Also, each switching element is considered as a three-terminal device that is reminiscent of a typical transistor. This basic model of each switching element became known as a switch-cell. The switch-cell terminal voltages and currents are averaged over each switching cycle. This averaging step is performed to validate the assumption of piece-wise linear outputs. However, the assumption of piece-wise input/output relations can reduce the bandwidth of the switch-cell and make it valid over a narrow range of switching frequencies.

The operational data models are developed based on the fact that inverters are considered parts of complete systems in different industrial applications. One of the famous examples of such an application is the utilization of renewable energy, where an inverter is considered as one component of a complete system model. Moreover, the requirements imposed by renewable energy systems define the operating point of the employed inverter. As a result, the inverter is considered as a single element, where input and output powers, voltages, and currents are taken as parameters of such an element. A transfer function relating input power with output power is developed based on the modeled system voltage-current relations. Furthermore, a curve-fit approximation is applied on collected inverter powers, voltages, and currents data to define a set of mathematical equations describing the final model.

The unified converter theory is based on considering switching converters related by their functions and behaviors. Also, the basic characteristics of switching converters depend neither on their applications, nor on their topologies. According to this theory, a typical switching converter is simply a matrix of switching elements that connects its input nodes to its output nodes. These input and output nodes can be ac or dc, capacitive or inductive. Moreover, the direction of the power flow can be from input nodes to output nodes or vice versa. This model is valid under constraints imposed by fundamental concepts of circuit theory, which include:

1. If one set of nodes (input or output) is inductive, the other set must be capacitive to avoid creating any cut-set of voltage or current sources when converter switches are activated.

2. Any combination of open and closed switches should never open circuit an inductor, or short circuit a capacitor.

Conventional inverter models are based on major assumptions of linearity and time-averaged switching actions. These assumptions have resulted in approximated and inaccurate models that cannot be generalized for modern pulse-width modulated inverters. Also, if these models are valid for certain operating conditions, they are not capable of justifying changes in switching technique or switching frequency. The assumption of time-averaged switching actions can be avoided if instantaneous switching actions are considered as part of the inverter model. One of the possible ways to validate this approach is using samplingreconstruction concepts.

## 3.2 SAMPLING-BASED MODELING OF SINGLE-PHASE INVERTERS

Multiswitching techniques are very common in operating and controlling modulated inverters. Such techniques are able to improve the performance of inverters in terms of output quality, efficiency, and dc-bus utilization. The fundamental idea of most modulation techniques is to compare a high-frequency signal known as the carrier (e.g., a triangular signal with frequency $f_c$) to a low-frequency signal known as the reference-modulating signal (usually a sinusoidal signal with frequency $f_m$). Also, some modulation techniques predefine switching instants like SHE and SVM. The reference-modulating signal has the same frequency as the desired output of any modulated inverter. Pulse-width modulated (PWM) and delta modulated (DM) inverters are very popular in different industrial applications. The harmonic spectra of outputs of PWM and DM inverters are shown in Figures 3.1 and 3.2, respectively. Note that the per-unit ($p.u$) in both figures represents the ratio of the magnitude to

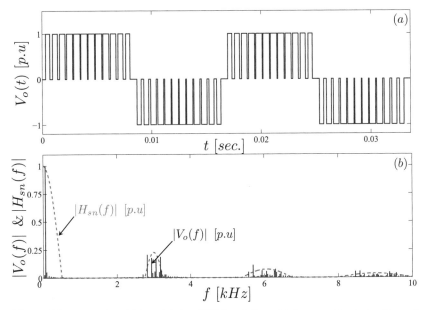

**Figure 3.1**    The pulse-width modulation (PWM) technique: (a) the per-unit output voltage of a single-phase $H$-bridge PWM inverter and (b) the harmonic spectrum of the output voltage.

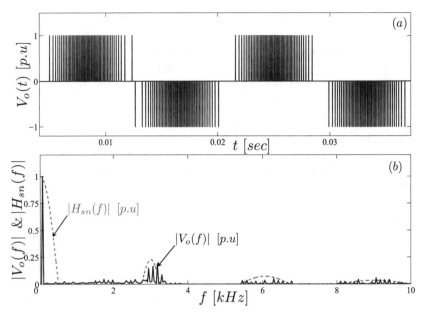

**Figure 3.2** The delta modulation (DM) technique: (a) the per-unit output voltage of a single-phase *H*-bridge DM inverter and (b) the harmonic spectrum of the output voltage.

a predefined value, which in these figures is taken as the maximum value of each quantity.

Figures 3.1 and 3.2 show harmonic components forming frequency side-bands centered at even multiples of the carrier frequency $f_c$. Similar frequency formations are found in spectra of reconstructed continuous-time (CT) signals from their samples.

## 3.2.1 Nonuniform Sampling-Based Representation

In general, one of the objectives of a modulation technique is to create trains of switching pulses that have the same volt-second average as a reference-modulating signal. The average is determined over a half cycle of the reference-modulating signal such as:

$$\int_0^{\frac{T_m}{2}} \left( \sum_L ON_L(t) \right) dt = \int_0^{\frac{T_m}{2}} S_M(t) dt \qquad (3.1)$$

where $T_m$ is the period of the reference-modulating signal $S_M(t)$ and the term $(\Sigma_L ON_L(t))$ is a train of ON-switching pulses during a half cycle of $S_M(t)$. The change in the status of the switching elements occurs at intersection points between the carrier signal and the reference-modulating signal. Unit impulses created at each intersection point can be viewed as nonuniform samples of the reference-modulating signal. Furthermore, each cycle of the carrier signal produces two samples; the rising portion of the carrier signal produces one sample, while the falling portion of the carrier signal produces the other sample. As a result, these samples appear to be taken at a sampling frequency of $2f_c$.

Samples created at intersection points have a nonuniform repetitive nature due to the periodicity and the symmetry of the carrier and the reference-modulating signals. This type of nonuniform sampling is known as the nonuniform recurrent sampling. The nonuniform recurrent sampling structure is based on arranging non-uniform samples into a limited number of repetitive groups, where each group has a finite number of samples. For the case of sampling a sinusoidal reference-modulating signal using a triangular carrier signal, each sample group will have two samples. The number of sample groups in one cycle of the sinusoidal reference-modulating signal will depend on the frequency of the reference-modulating signal and the frequency of the carrier signal. If the sinusoidal reference-modulating signal has a frequency of $f_m$ and the carrier signal has a frequency of $f_c$, the number of sample groups can be defined as:

$$\mathcal{D} = 2\frac{f_c}{f_m} \tag{3.2}$$

The formed groups of nonuniform recurrent samples over one cycle of the reference-modulating signal can be viewed as a set $\mathcal{D} = \{d_1, d_2, \ldots, d_{\mathcal{D}}\}$. The created nonuniform recurrent sample groups can take a discrete form of the reference-modulating signal $S_M(t)$ that can be expressed as:

$$S_{dM}(t) = \sum_{k=-\infty}^{\infty} \sum_{d=0}^{\mathcal{D}} \sum_{p=1}^{2} S_M(t)\delta\left(t - t_{pd} - kT_m\right) \tag{3.3}$$

where the variable $t_{pd}$ represents $d$ groups of $p$ samples, which form one of the recurrent periods. The sampled form of the reference-modulating signal can be expressed as a discrete signal as:

$$S_{dM}[n] = S_{dM}(t)\big|_{t=t_{pd}+kT_m} \tag{3.4}$$

This discrete form provides a basis for viewing the instantaneous switching actions of an inverter as stages of interpolating functions. Such interpolating functions are used to recover $S_M(t)$ from its samples $Sd_M[n]$.

The popular sinusoidal pulse-width modulation (SPWM) technique uses a triangular carrier signal for creating trains of switching pulses. These pulses are characterized by their widths and locations relative to the beginning of each cycle of $S_M(t)$. In SPWM, nonuniform recurrent samples are created by the carrier signal that can be expressed mathematically as:

$$S_{SC}(t) = \begin{cases} 4f_c t & 0 \leq t \leq \dfrac{T_c}{4} \\[2mm] 2 - 4f_c t & \dfrac{T_c}{4} \leq t \leq \dfrac{3T_c}{4} \\[2mm] 4f_c t - 4 & \dfrac{3T_c}{4} \leq t \leq T_c \end{cases} \tag{3.5}$$

Two signals of different functional forms representing the rising and the falling portions of $S_{SC}(t)$ that are responsible for creating two samples for each sample group, can be defined as trains of nonuniform recurrent impulses as:

$$S_{SCR}(t) = \sum_{k} \sum_{d=1}^{D} \delta(t - t_{Rdr} - dT_c - kT_m) \tag{3.6}$$

$$S_{SCF}(t) = \sum_{k} \sum_{d=1}^{D} \delta(t - t_{Fdr} - dT_c - kT_m) \tag{3.7}$$

The discrete form of these two signals can be defined as:

$$S_{SCR}[n] = S_{SCR}(t)\big|_{t=t_{Rdr}-dT_c-kT_m} \tag{3.8}$$

$$S_{SCF}[n] = S_{SCF}(t)\big|_{t=t_{Fdr}-dT_c-kT_m} \tag{3.9}$$

where the set $\{t_{Rdr}\}$ represents the intersection points of the rising portion of $S_{SC}(t)$ with $S_M(t)$, while the set $\{t_{Fdr}\}$ represents the intersection points of the falling portion of $S_{SC}(t)$ with $S_M(t)$. These sets of time instants can be determined over a cycle of $S_{SC}(t)$ as:

$$S_M(t_{Rr}) - 4f_c t_{Rr} = 0 \tag{3.10}$$

$$S_M(t_{Fr}) - (2 - 4f_c t_{Fr}) = 0 \tag{3.11}$$

Figure 3.3 shows the intersection points of both discrete signals $S_{SCR}[n]$ and $S_{SCF}[n]$ with a sinusoidal reference-modulating signal.

The period of sample groups $T_m$ is related to the Nyquist interval $T_Q$ and the number of sample groups $\mathcal{D}$ by the following relation:

$$T_m > N\mathcal{D}T_Q \tag{3.12}$$

where $N$ is the number of samples in each sample group, which is $N = 2$ for the SPWM technique case. Figure 3.4 shows the discrete form of the reference-modulating signal $S_M(t)$.

The discrete signal $S_{SCR}[n]$ is the result of sampling the reference-modulating signal $S_M(t)$ with the rising portion of the carrier signal, while the discrete signal $S_{SCF}[n]$ is the result of sampling $S_M(t)$ with the falling portion. This interpretation of creating samples of the reference-modulating signal $S_M(t)$ can be constructed in a block diagram as shown in Figure 3.5.

### 3.2.2  Reconstructing the Reference-Modulating Signal from Nonuniform Samples

The reconstruction of a continuous-time (CT) signal from its samples is carried out through a filtering process. The classical sampling theorem states that if a CT signal $s(t)$ has a Fourier transform $F\{s(t)\} = S(\Omega) = 0$ for $|\Omega| > \Omega_o$ (i.e., the signal $s(t)$ is

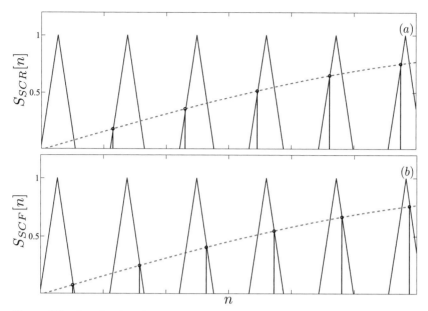

**Figure 3.3**    Sampling the reference-modulating signal $S_M(t)$ in a nonuniform recurrent manner: (a) $S_{SCR}[n]$; the samples created by the rising portion of $S_{SC}(t)$ and (b) $S_{SCF}[n]$; the samples created by the falling portion of $S_{SC}(t)$.

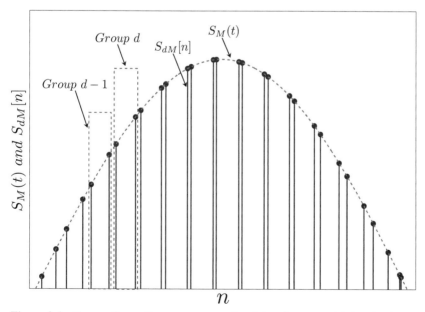

**Figure 3.4**    Groups of nonuniform recurrent samples of the reference-modulating signal $S_M(t)$ using both discrete signals $S_{SCR}(t)$ and $S_{SCF}(t)$ to form $S_{dM}[n]$.

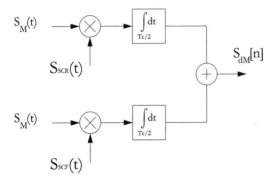

**Figure 3.5** A block diagram of nonuniform sampling the reference-modulating signal using the rising portion $S_{SCR}[n]$ and the falling portion $S_{SCF}[n]$ of the triangular carrier signal $S_{SC}(t)$.

band-limited to $\Omega_o$), then $s(t)$ can be recovered perfectly from its samples $s_d[n]$ by the following formula:

$$s(t) = \sum_{n=-\infty}^{\infty} s_d[n]\frac{\sin(\pi(t-n))}{\pi(t-n)} \tag{3.13}$$

The reconstruction formula of equation (3.13) represents an interpolating function, which is valid under a strict condition of equispaced samples. However, for the case of nonuniform recurrent sampling, reconstructing the CT signal is carried out through stages of interpolating functions. This can be realized using different stages of filters or a filter bank, where the number of the required filters depends on the number of sample groups $\mathcal{D}$. In general, a CT signal $x_c(t)$ can be reconstructed from its nonuniform recurrent samples $x_d[n]$ using the Lagrange general interpolating formula as:

$$x_c(t) = \sum_{n=-\infty}^{\infty} x_d[n]\frac{G_n(t)}{G_n'(t_n)(t-t_n)} \tag{3.14}$$

where

$$G_n(t) = t\prod_p\left(1-\frac{t}{t_p}\right) \tag{3.15}$$

and

$$G_n'(t_p) = \frac{dG_n(t)}{dt}\bigg|_{t=t_p} \tag{3.16}$$

For the case of $N = 2$, the function $G_n(t)$ can be simplified to:

$$G_n(t) = t\left(1-\frac{t}{t_1}\right)\left(1-\frac{t}{t_2}\right) \tag{3.17}$$

The case of $N = 2$ can be employed to express the Lagrange interpolating function for the SPWM technique so that $t_1 = t_{Rdr}$ and $t_2 = t_{Fdr}$. Hence, the Lagrange interpolation formula can be expressed for the case of SPWM technique ($N = 2$) as:

$$S_{MS}(t) = \sum_r \sum_{d=1}^{D} \sum_{p=1}^{2} S_{dM}[pd] \frac{G_d(t)}{G_d'(t_p)(t - t_p)} \tag{3.18}$$

One group of nonuniform recurrent samples is created for each cycle of the triangular carrier signal $S_{SC}(t)$. Also, one stage of Lagrange interpolating functions is defined over that group of samples. This representation can be stated as:

$$\{ S_M(t_{d1}), S_M(t_{d2}) \}, t \in [t_{d1}, t_{d2}] \tag{3.19}$$

where $\{ S_M(t_{d1}), S_M(t_{d2}) \}$ is the sample group $d$. The interpolating function for the nonuniform recurrent sample group $d$ can be defined as:

$$\lambda_d(t) = \frac{S_M(t_{d1}) G_d(t)}{G_d'(t_{d1})(t - t_{d1})} + \frac{S_M(t_{d2}) G_d(t)}{G_d'(t_{d2})(t - t_{d2})} \tag{3.20}$$

where the function $G_d(t)$ is defined for the sample group $d$ as:

$$G_d(t) = t \left( 1 - \frac{t}{t_{d1}} \right) \left( 1 - \frac{t}{t_{d2}} \right), t \in [t_{d1}, t_{d2}] \tag{3.21}$$

The time interval $[t_{d1}, t_{d2}]$ is known as the interval of support of the interpolating function $\lambda_d(t)$ over the sample group $d$. Due to the periodicity of the sample groups, the function $\lambda_d(t)$ is periodic with a period of $T_m$ so that:

$$\lambda_d(t) = \begin{cases} \lambda_d(t - rT_m) & t_{d1} \leq t \leq t_{d2} \\ 0 & \text{otherwise} \end{cases} \tag{3.22}$$

where $r = 0, 1, 2, \dots$. Figure 3.6 shows two successive interpolating functions and their associated inverter output switching actions.

Successive interpolating functions $\{\lambda_d(t)\}_{d=1,2\dots,D}$ are defined through simplifying the Lagrange general interpolating formula as given in equation (3.17) for the case of the inverter that is characterized by $N = 2$. Moreover, the interval of support for each interpolating function $\lambda_d(t)$ is stated as $\{[td_1, td_2]\}_{d=1,2,\dots,D}$. These interpolation functions can produce a reconstructed CT signal to model the output of a SPWM inverter $V_o(t)$ as:

$$V_o(t) = V_{DC} \sum_{r=0}^{\infty} \sum_{d=1}^{D} \lambda_d(t - rT_m) \tag{3.23}$$

Figure 3.7 shows two cycles of the reconstructed $V_o(t)$ along with its associated inverter output voltage. The two reconstructed signals are analyzed using Fourier analysis, the harmonic distributions for both of them are shown in Figure 3.8.

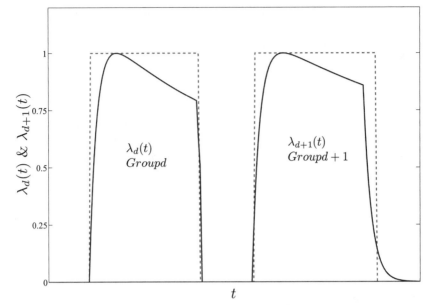

**Figure 3.6**   Two successive interpolating functions $\lambda_d(t)$ and $\lambda_{d+1}(t)$ and their normalized associated ON-switched inverter output.

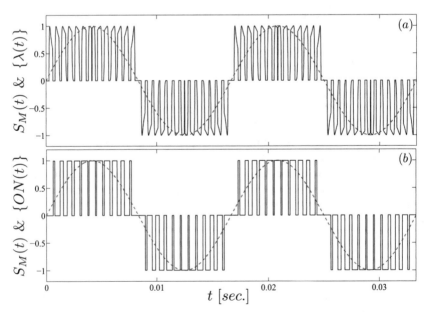

**Figure 3.7**   Reconstructing sinusoidal signals for the PWM case: (a) reconstructing a sinusoidal signal using the interpolating functions generated by the sampling-based model ($\{\lambda_d(t)\}_{d=1,2...,\mathcal{D}}$) and (b) the reconstructed sinusoidal signal using switching pulses generated by the PWM technique.

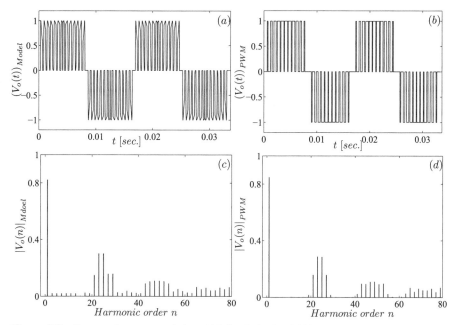

**Figure 3.8**   Spectra of reconstructed sinusoidal signals for $f_c = 1.08\,kHz$: (a) the reconstructed sinusoidal signal using $\{\lambda_d(t)\}_{d=1,2,...,\mathcal{D}}$, (b) the reconstructed sinusoidal signal using switching pulses generated by the SPWM technique, (c) the spectrum of the reconstructed sinusoidal signal using $\{\lambda_d(t)\}_{d=1,2,...,\mathcal{D}}$, and (d) the spectrum of the reconstructed sinusoidal signal using switching pulses generated by the SPWM technique. $n$ is the harmonic order.

Instantaneous switching actions are considered a main part of the developed inverter sampling-based model. Such consideration makes this model capable of simulating the performance of modulated inverters. As Figure 3.8 shows, the outputs of the developed model are almost identical to the actual inverter outputs. The nonuniform recurrent sampling-based model is tested for different operating conditions that include changing the switching frequency. The next section provides the simulation test results for these cases.

## 3.3   TESTING THE NONUNIFORM RECURRENT SAMPLING-BASED MODEL OF INVERTERS

The developed nonuniform recurrent sampling-based model represents the inverter output as a reconstructed CT signal using sets of interpolating functions. The proposed model is tested for producing the output voltage of a SPWM inverter for two carrier frequencies.

### 3.3.1 PWM Inverter Output Voltage for Two Carrier Frequencies

In general, the carrier frequency of a PWM technique determines the rate at which inverter switching elements change their status. If the carrier frequency $f_c$ is changed, locations as well as the number of intersection points of $S_{SC}(t)$ with $S_M(t)$ will change. Moreover, changing $f_c$ has a direct impact on the spectral distribution of the SPWM inverter output harmonic components. From the perspective of the proposed model, changing $f_c$ affects locations of samples as well as the number of sample groups created over each cycle of $S_M(t)$. As a result, intervals of support and locations of the proposed model interpolating functions are affected. Figures 3.9(a) and 3.9(b) show successive nonuniform recurrent sample groups of $S_M(t)$ with $f_m = 60\,Hz$ for two common values of switching frequency $f_c = 1.08\,kHz$ and $1.8\,kHz$, respectively.

Figure 3.10 shows the normalized output of a PWM inverter for a carrier frequency of $f_c = 1.8\,kHz$ and sets of interpolating functions produced by the proposed sampling model for this value of $f_c$ along with their spectra. The reconstructed CT signal using model sets of interpolating functions has a spectrum that is very close to the output voltage of a SPWM inverter, which confirms its accuracy.

The nonuniform sampling model is based on considering instantaneous switching actions rather than averaging them over time. This main feature of the proposed model has made it possible to verify the effects on the inverter output due to dif-

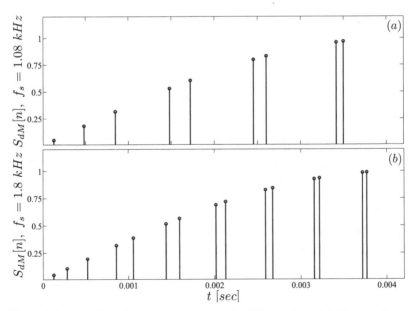

**Figure 3.9**   Nonuniform recurrent sample groups for different values of $f_c$: (a) successive sample groups of $S_M(t)$ for $f_c = 1.08\,kHz$ and (b) successive sample groups of $S_M(t)$ for $f_c = 1.8\,kHz$.

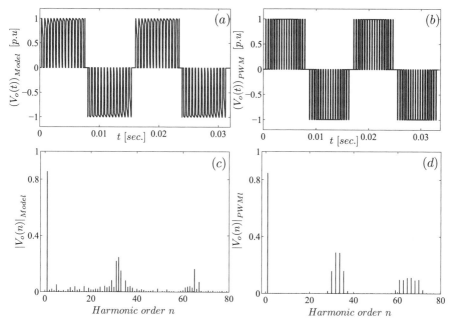

**Figure 3.10**    Spectra of reconstructed sinusoidal signals for a carrier frequency of $f_c = 1.8\,kHz$: (a) the reconstructed sinusoidal signal using $\{\lambda_d(t)\}_{d=1,2...,\mathcal{D}}$, (b) the reconstructed sinusoidal signal using switching pulses generated by the SPWM technique, (c) the spectrum of the reconstructed sinusoidal signal using $\{\lambda_d(t)\}_{d=1,2...,\mathcal{D}}$, and (d) the spectrum of the reconstructed sinusoidal signal using switching pulses generated by the SPWM technique. $n$ id the harmonic order.

ferent modifications of switching strategy. It is worth mentioning that existing inverter models lack the adequate capability to interpret impacts on inverter outputs due to any change in the switching strategies involving nonsinusoidal reference modulating signals and multiple carrier frequencies. The proposed nonuniform recurrent sampling model can easily create the outputs and their harmonic spectra of an inverter. If spacings between samples are selected so that optimal interpolating functions can be used, then a perfect reconstruction of the sinusoidal reference-modulating signal becomes possible. Such optimization can be achieved using signal processing and wavelets concepts. It is worth mentioning that the proposed nonuniform recurrent sampling-based model has been used to develop new carrier signals and new modulation technique. These are used to improve the performance of power inverters.

## 3.4  SAMPLING-BASED MODELING OF THREE-PHASE INVERTERS

The nonuniform recurrent sampling-based mathematical model of $1\phi$ inverters has shown an encouraging accuracy for modeling the inverter 's function. This approach

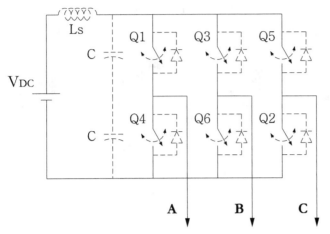

**Figure 3.11**   Schematic diagram of a typical 3φ three-leg, six-pulse inverter.

of modeling single-phase inverters can be extended for 3φ inverters. In a typical 3φ inverter, three reference-modulating signals are used to generate the required switching pulses. These three signals are shifted $\dfrac{2\pi}{3}$ *radian* from each other so that each one of them is associated with one phase on the output side of the 3φ inverter. Note that 3φ inverters can have different configurations. However, the three-leg, six-pulse configuration is the most common one in industrial applications. Figure 3.11 shows a schematic diagram of the common three-leg, six-pulse topology of 3φ inverters.

The required reference-modulating signals for a 3φ inverter are given by:

$$S_{Ma}(t) = \sin(\omega_m t) \tag{3.24}$$

$$S_{Mb}(t) = \sin\left(\omega_m t - \frac{2\pi}{3}\right) \tag{3.25}$$

$$S_{Mc}(t) = \sin\left(\omega_m t + \frac{2\pi}{3}\right) \tag{3.26}$$

The carrier-based techniques are the most common techniques used to generate switching pulses for 1φ as well as 3φ inverters. In theses techniques, switching pulses widths and locations are determined by the locations of intersection points of reference-modulating signals and a carrier signal. Triangular signals are widely used in carrier-based techniques to generate trains of periodic switching pulses. This process can be viewed as multiplying each reference-modulating signal with a train of impulses located at the intersection points. For the case of 3φ inverters, three trains of impulses are required to generate switching pulses for each leg of the $3\phi$ inverter. These trains of impulses can be expressed as:

$$P_a(t) = \sum_r \sum_{d=1}^{D} \sum_{pa=1}^{2} \delta(t - t_{pa} - dT_c - rT_m) \tag{3.27}$$

$$P_b(t) = \sum_r \sum_{d=1}^{D} \sum_{pb=1}^{2} \delta(t - t_{pb} - dT_c - rT_m) \tag{3.28}$$

$$P_c(t) = \sum_r \sum_{d=1}^{D} \sum_{pc=1}^{2} \delta(t - t_{pc} - dT_c - rT_m) \tag{3.29}$$

where $T_c = \dfrac{1}{f_c}$ is the period of the carrier signal, $T_m = \dfrac{1}{f_m}$ is the period of each reference-modulating signal and $\delta(t)$ is the Dirac delta function. Using these trains of impulses, the three reference-modulating signals can be sampled in a nonuniform recurrent manner. The sampled versions of these reference-modulating signals can be stated as:

$$S_{dM}[n]_a = \int_{\frac{T_c}{2}} S_{Ma}(t) P_a(t) \, dt \tag{3.30}$$

$$S_{dM}[n]_b = \int_{\frac{T_c}{2}} S_{Mb}(t) P_b(t) \, dt \tag{3.31}$$

$$S_{dM}[n]_c = \int_{\frac{T_c}{2}} S_{Mc}(t) P_c(t) \, dt \tag{3.32}$$

Figure 3.12 shows the three reference-modulating signals, the triangular carrier signal, and the resultant nonuniform recurrent sampled versions of the three reference-modulating signals.

Each cycle of the carrier signal produces two samples for each sinusoidal reference-modulating signal. The time intervals $[t_{ad1}, t_{ad2}]$, $[t_{bd1}, t_{bd2}]$, and $[t_{cd1}, t_{cd2}]$ are intervals of the group $d$ of nonuniform recurrent samples of $S_{Ma}(t)$, $S_{Mb}(t)$, and $S_{Mc}(t)$, respectively. This representation is considered a line-to-line one due to the operation of a typical $3\phi$ three-leg, six-pulse inverter. In $3\phi$ voltage-source (VS) six-pulse inverters, the voltage across each leg is switched from $+V_{DC}$ to $-V_{DC}$ alternately. Such an alternate switching causes each leg to change its connection with the other inverter legs. A leg appears in series with a parallel connection of the other two legs; when switched, it becomes in parallel with one leg and both are in series with the third leg. The changes in the connections over a time interval of $T_m$ are illustrated in Figure 3.13. A similar alternate switching takes place in $3\phi$ current-source (CS) six-pulse inverters. The dc current flowing through each leg is switched alternately causing inverter legs to undergo similar alternate connections.

The alternate switching of a $3\phi$ inverter produces outputs that can be measured either as line-to-line or as line-to-neutral quantities. Figure 3.14 shows line-to-line and line-to-neutral quantities for a square wave switching operation.

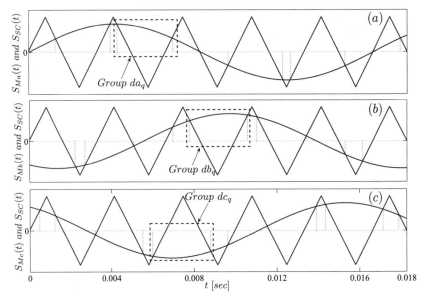

**Figure 3.12**   Nonuniform sample groups of the three reference-modulating signals $S_{Ma}(t)$, $S_{Mb}(t)$, and $S_{Mc}(t)$ created using the same triangular carrier signal $S_{SC}(t)$.

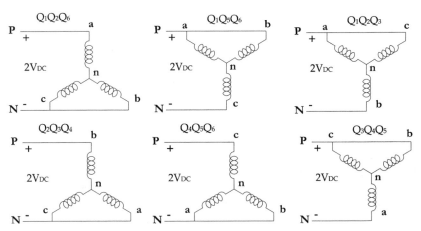

**Figure 3.13**   Inverter legs connection changes due to alternate switching of the DC supply.

When a $3\phi$ six-pulse inverter is switched through a modulation process, both line-to-line and line-to-neutral quantities appear as trains of ON switching pulses. Such outputs are shown in Figure 3.15.

The previous description of a $3\phi$ inverter operation indicates that switching signals are generated based on a line-to-line quantity for each leg. This result can be utilized in developing a sampling-based model of a $3\phi$ inverter. As the three

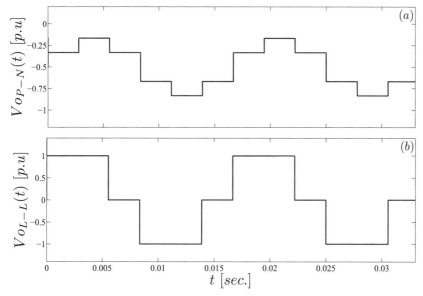

**Figure 3.14**    $3\phi$ six-pulse inverter output voltages: (a) the line-to-neutral voltage ($V_{oP-N}$) for one phase leg and (b) the output line-to-line voltage ($V_{oL-L}$) for the same phase leg. The base value is $V_{oL-L}$.

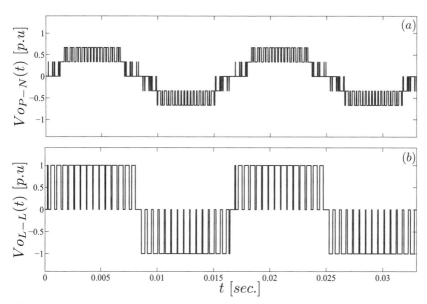

**Figure 3.15**    $3\phi$ PWM six-pulse inverter output voltages: (a) per unit line-to-neutral voltage ($Vo_{P-N}$) and (b) per unit line-to-line voltage ($V_{oL-L}$). The base value is $V_{oL-L}$.

reference-modulating signals are sampled in a nonuniform recurrent manner, the reconstruction can be carried out as three independent reconstruction processes. As a consequence, three switches located in different legs of the $3\phi$ inverter are activated at any given time. One of these switches is replaced by another switch, each $\alpha_\mu$, where $\alpha_\mu$ is expressed as:

$$\alpha_\mu = \mu\omega_m \frac{T_m}{12}; \mu = 1, 2, \ldots, 12 \tag{3.33}$$

where $\omega_m = 2\pi f_m$. Each switching element is activated for a period of $\dfrac{T_m}{2}$ seconds.

This is necessary to avoid creating any short circuits in parallel with any inverter leg. The reconstruction of the three reference-modulating signals from their nonuniform recurrent samples is carried out using Lagrange interpolation functions, which can be defined for sample group $d$ for $ab\ c$ phases as:

$$\lambda_{ad}(t) = \frac{S_{Ma}(t_{ad1})G_{ad}(t)}{G'_{ad}(t_{ad1})(t - t_{ad1})} + \frac{S_{Ma}(t_{ad2})G_{ad}(t)}{G'_{ad}(t_{ad2})(t - t_{ad2})} \tag{3.34}$$

$$\lambda_{bd}(t) = \frac{S_{Mb}(t_{bd1})G_{bd}(t)}{G'_{bd}(t_{bd1})(t - t_{bd1})} + \frac{S_{Mb}(t_{bd2})G_{bd}(t)}{G'_{bd}(t_{bd2})(t - t_{bd2})} \tag{3.35}$$

$$\lambda_{cd}(t) = \frac{S_{Mc}(t_{cd1})G_{cd}(t)}{G'_{cd}(t_{cd1})(t - t_{cd1})} + \frac{S_{Mc}(t_{cd2})G_{cd}(t)}{G'_{cd}(t_{cd2})(t - t_{cd2})} \tag{3.36}$$

where $G_{ad}(t)$, $G_{bd}(t)$, and $G_{cd}(t)$ are Lagrangian interpolating functions over the group $d$ of nonuniform recurrent samples, and are given for $abc$ phases by:

$$G_{ad}(t) = t \prod_p \left(1 - \frac{t}{t_{adp}}\right) \tag{3.37}$$

$$G_{bd}(t) = t \prod_p \left(1 - \frac{t}{t_{bdp}}\right) \tag{3.38}$$

$$G_{cd}(t) = t \prod_p \left(1 - \frac{t}{t_{cdp}}\right) \tag{3.39}$$

Also, $G'_{ad}(t)$, $G'_{bd}(t)$, and $G'_{cd}(t)$ are the first derivatives of the Lagrangian interpolating functions over the sample group $d$ that can be defined for $abc$ phases as:

$$G'_{ad}(t_{adp}) = \left. \frac{dG_{ad}(t)}{dt} \right|_{t=t_{adp}} \tag{3.40}$$

$$G'_{bd}(t_{bdp}) = \left. \frac{dG_{bd}(t)}{dt} \right|_{t=t_{bdp}} \tag{3.41}$$

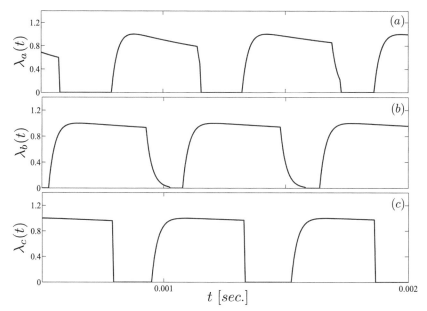

**Figure 3.16**   The three interpolation function for two adjacent groups of samples: (a) the interpolation function for phase $a$ leg: $\lambda_a(t)$, (b) the interpolation function for phase $b$ leg: $\lambda_b(t)$, and (c) the interpolation function for phase $c$ leg: $\lambda_c(t)$.

$$G'_{cd}\left(t_{cdp}\right) = \left.\frac{dG_{cd}\left(t\right)}{dt}\right|_{t=t_{cdp}} \tag{3.42}$$

where $p = 1, 2$. Time intervals $[t_{ad1}, t_{ad2}]$, $[t_{bd1}, t_{bd2}]$, and $[t_{cd1}, t_{cd2}]$ are time intervals for the three interpolating functions $\lambda_{ad}(t)$, $\lambda_{bd}(t)$, and $\lambda_{cd}(t)$ for the samples groups $ad$, $bd$, and $cd$, respectively. Due to the periodicity of sample groups, these three interpolating functions are periodic with a period of $T_m$, and can be defined for sample groups $ad$, $bd$, and $cd$ as follows:

$$\lambda_{ad}\left(t\right) = \begin{cases} \lambda_{ad}\left(t - rT_m\right) & t_{ad1} \leq t \leq t_{ad2}, r \in \mathbb{Z} \\ 0 & \text{otherwise} \end{cases} \tag{3.43}$$

$$\lambda_{bd}\left(t\right) = \begin{cases} \lambda_{bd}\left(t - rT_m\right) & t_{bd1} \leq t \leq t_{bd2}, r \in \mathbb{Z} \\ 0 & \text{otherwise} \end{cases} \tag{3.44}$$

$$\lambda_{cd}\left(t\right) = \begin{cases} \lambda_{cd}\left(t - rT_m\right) & t_{cd1} \leq t \leq t_{cd2}, r \in \mathbb{Z} \\ 0 & \text{otherwise} \end{cases} \tag{3.45}$$

where $d = 1, 2, \ldots, \mathcal{D}$. Note that $\mathcal{D}$ is the number of sample groups over one cycle of each reference-modulating signal. Figure 3.16 shows $\lambda_a(t)$, $\lambda_b(t)$ and $\lambda_c(t)$

evaluated for two adjacent sample groups $d$ and $d + 1$ for each reference-modulating signal.

The definition of the three interpolating functions leads to stating the sampling-based model of a $3\phi$ six-pulse inverter as:

$$V_{ab}(t) = V_{DC}\sum_{r}\sum_{d=1}^{D}\lambda_{ad}(t - rT_m) \tag{3.46}$$

$$V_{bc}(t) = V_{DC}\sum_{r}\sum_{d=1}^{D}\lambda_{bd}(t - rT_m) \tag{3.47}$$

$$V_{ca}(t) = V_{DC}\sum_{r}\sum_{d=1}^{D}\lambda_{cd}(t - rT_m) \tag{3.48}$$

Line-to-neutral output voltages can be derived from line-to-line ones as:

$$V_{an}(t) = \frac{V_{ab}\left(t - \frac{T_m}{12}\right)}{\sqrt{3}} \tag{3.49}$$

$$V_{bn}(t) = \frac{V_{bc}\left(t - \frac{T_m}{12}\right)}{\sqrt{3}} \tag{3.50}$$

$$V_{cn}(t) = \frac{V_{ca}\left(t - \frac{T_m}{12}\right)}{\sqrt{3}} \tag{3.51}$$

Figure 3.17 shows the three sets of interpolating functions used to reconstruct $3\phi$ line-to-line output voltages for a $3\phi$ VS PWM six-pulse inverter. Figure 3.18 shows the three sets of interpolating functions used to reconstruct $3\phi$ line-to-neutral output voltages for a $3\phi$ VS PWM six-pulse inverter.

The developed nonuniform recurrent sampling-based approach is used to model a $3\phi$ six-pulse inverter. Furthermore, the input data to this model includes $f_m$, $f_c$ and the set of intersection points over one cycle of each reference-modulating signal $\{(t_{dp})_a\}$ $\{(t_{dp})_b\}$ and $\{(t_{dp})_c\}$, where $p = 1, 2, d = 1, 2,..., D$:

$$d = q + 1 \tag{3.52}$$

Phase A normalized line-to-line $V_{AB}(t)$ and line-to-neutral $V_{AN}(t)$ voltages obtained using the proposed sampling-based model are compared with phase A voltages obtained on the output of a $3\phi$ six-pulse VS PWM inverter. Figure 3.19 shows inverter actual phase A voltages along with their spectra and phase A voltages obtained using the sampling-based model along with their spectra.

As shown in Figure 3.19, phase A voltages obtained using the sampling-based model are almost identical to the actual voltages. Also, the spectra of both types of

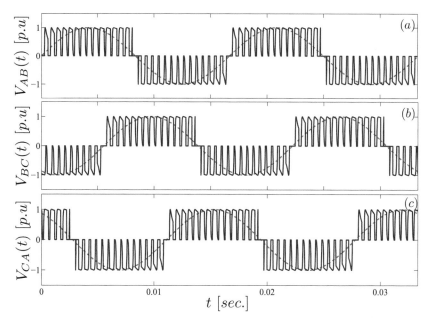

**Figure 3.17**    Reconstructing $3\phi$ line-to-line output voltages for a $3\phi$ VS PWM six-pulse inverter:
(a) output of phase $a$ leg: $V_{AB}(t)$, (b) output of phase $b$ leg: $V_{BC}(t)$, and (c) output of phase $c$ leg: $V_{CA}(t)$.

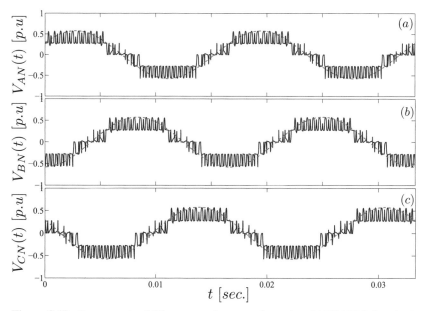

**Figure 3.18**    Reconstructing $3\phi$ line-to-neutral output voltages for a $3\phi$ VS PWM six-pulse inverter:
(a) output of phase $a$: $V_{AN}(t)$, (b) output of phase $b$: $V_{BN}(t)$, and (c) output of phase $c$: $V_{CN}(t)$.

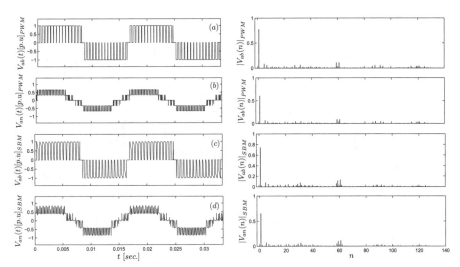

**Figure 3.19**    The output voltage of $3\phi$ VS six-pulse inverter: (a) the output voltage of phase $a$ leg: $V_{ab}(t)$ (actual inverter output) along with its spectrum, (b) the output voltage of phase $a$ leg: $V_{an}(t)$ (actual inverter output) along with its spectrum, (c) the output voltage of phase $a$ leg: $V_{ab}(t)$ using interpolating functions $\{\lambda_d(t)\}_{d=1,2,...,\mathcal{D}}$ along with its spectrum, and (d) the output voltage of phase $a$ leg: $V_{an}(t)$ using interpolating functions $\{\lambda_d(t)\}_{d=1,2,...,\mathcal{D}}$ along with its spectrum.

voltages are very close, which demonstrates the accuracy of the proposed nonuniform recurrent sampling-based model of $3\phi$ inverters.

## 3.5 SUMMARY

This chapter has provided a brief review of available inverter models along with their structures, conditions for validation, limitations, and relations with switching techniques. Several modulation techniques have been developed and tested to improve the performance of inverters for different applications. In addition, inverter modulation techniques have always been optimized with the objective of determining the instants at which inverter's switching element change their status (ON to OFF or OFF to ON). This objective can be viewed as determining the switching instants so that a reconstructed inverter output will have most of its energy concentrated in one frequency. Moreover, the reconstruction of the inverter output, given the switching instants, is realized by trains of rectangular pulses. Such a process can be better optimized through selecting the switching instants based on the nature of the reconstruction functions. This approach can be accurately implemented using concepts of the sampling theorem and the wavelet theory. The sampling-based modeling can be utilized without any need for a carrier signal or predefined switching angles so that the inverter function is mathematically characterized.

  A new approach based on concepts from the sampling theorem has been incorporated to construct models for $1\phi$ and $3\phi$ inverters. The main concept from the sampling theorem has been the nonuniform recurrent sampling-reconstruction of CT signals. The connection between the sampling theorem and the wavelet theory can be utilized to develop an ideal sampling-reconstruction process for operating any inverter to achieve an optimal performance. The next chapter provides a method for optimizing the nonuniform recurrent sampling-reconstruction of CT signals using new wavelet basis functions that can be employed for switching $1\phi$ and $3\phi$ inverters.

# Chapter 4

# Scale-Based Linearly Combined Wavelets

## 4.1 INTRODUCTION

The preceding chapter presented a new model for both single-phase ($1\phi$) and three-phase ($3\phi$) inverters based on the nonuniform sampling-reconstruction of continuous-time (CT) signals. Chapter 2 demonstrated that sampling a CT signal can be represented as an $N$-dimensional approximation case. Such a representation has been used to interpret the sampling theorem in the context of the wavelet-based multi-resolution analysis (MRA). Moreover, Chapter 2 provided a brief review of different types of wavelet basis functions and their associated MRAs. This chapter presents a new type of wavelet basis functions that are capable of constructing nondyadic-type MRAs, which can support nonuniform recurrent sampling-reconstruction structures.

The literature of wavelets provides several types of wavelet basis functions that have been employed for processing signals with complex time-frequency structures. Such signals are popular in various applications in signal and image processing areas. On the other hand, different wavelet basis functions are capable of constructing dyadic-type MRAs. These dyadic MRAs can support uniform sampling-reconstruction cases. In some applications of signal and image processing as well as switched power electronic converters, uniform sampling has been found to be redundant and may suffer from implementation problems. In such applications, nonuniform sampling has been found to be more practical for implementation. As a result, new MRA structures are required to support nonuniform sampling-reconstruction processes that include the nonuniform recurrent sampling. The next section presents a new type of wavelet basis functions that are capable of spanning spaces to construct a nondyadic-type MRA to support nonuniform recurrent sampling-reconstruction processes.

*An Introduction to Wavelet Modulated Inverters*, by S.A. Saleh and M. Azizur Rahman
Copyright © 2011 Institute of Electrical and Electronics Engineers

## 4.2  SCALE-BASED LINEARLY COMBINED WAVELET BASIS FUNCTIONS

Wavelet basis functions and their related transforms have been used as effective and efficient tools for representing and processing signals with nonperiodic and nonstationary natures, which are common in power systems, machines drives, and power electronic converters. These features of wavelet functions are based on their ability to parameterize signals using sets of basis functions that are localized in time and frequency. This type of signal processing involves decomposing signals using sets of analysis basis functions, and reconstructing signals using sets of synthesis basis functions as generalized in the following equation:

$$x(t) = \sum_{\alpha=0}^{\infty} \sum_{\beta=-\infty}^{\infty} \langle \xi_a(2^\alpha t - \beta), x(t) \rangle \xi_s(2^\alpha t - \beta) \tag{4.1}$$

where $\{\xi_a(2^\alpha t - \beta)\}_{\alpha,\beta \in \mathbb{Z}}$ and $\{\xi_s(2^\alpha t - \beta)\}_{\alpha,\beta \in \mathbb{Z}}$ are sets of wavelet analysis and synthesis basis functions, respectively. The analysis and synthesis of signals using wavelet basis functions is commonly defined as an MRA. The fundamental idea of constructing an MRA is to define a scale $j$ such that sets of basis functions can span a collection of complete and dense spaces. Such sets of basis functions are expressed as:

$$\{\{\phi_{j,k}(t)\}, \{\psi_{j,k}(t)\}\} = \{\{\phi(2^j t - k)\}, \{\psi(2^j t - k)\}\}, \quad j, k \in \mathbb{Z} \tag{4.2}$$

where $\phi(t)$ is a scaling function and $\psi(t)$ is a wavelet function. At each scale $j$, two spaces are spanned by one set of scaling basis functions $V_j(\phi)$ and one set of wavelet basis functions $W_j(\psi)$. These spaces can be related to the desired MRA as:

$$MRA(j) = V_j(\phi) \oplus W_j(\psi) \tag{4.3}$$

where $\oplus$ is the orthogonal sum operation. This structure of an MRA can be used to expand a signal $f(t)$ using sets of basis functions up to scale $j$ as:

$$f(t) = \sum_{k \in \mathbb{Z}} \langle f(t), \phi_{j,k}(t) \rangle \tilde{\phi}_{j,k}(t) + \sum_{j \in \mathbb{Z}} \sum_{k \in \mathbb{Z}} \langle f(t), \psi_{j,k}(t) \rangle \tilde{\psi}_{j,k}(t) \tag{4.4}$$

where $\tilde{\phi}_{j,k} = \tilde{\phi}(2^j t - k)$ is a synthesis scaling function and $\tilde{\psi}_{j,k} = \tilde{\psi}(2^j t - k)$ is a synthesis wavelet function.

In general, any MRA characteristic depends on the nature of its nested spaces ($V_j(\phi)$ and $W_j(\psi)$) that are spanned by sets of basis functions. The scaling function is considered as the key element for defining the nature of an MRA. MRAs constructed using wavelet basis functions have been successfully used in different applications including power quality improvements, detecting and classifying faults and transient disturbances in power systems and power electronics, analysis of power converters, and detection of peak voltages and current in different circuit topologies.

The focus of this book is on developing an MRA that is capable of supporting a nonuniform sampling theorem for switching a dc-ac inverter. An ON switching pulse for a typical inverter can be modeled using Haar scaling functions. Moreover, the duration and location of each ON switching pulse changes in a nondyadic manner. Such a sequence of nondyadic Haar scaling functions can be viewed as a type of scaling function. This can be realized by defining a scaling function as follows:

*Definition:* let $\varphi(t) = \varphi_1(t)$ be a *scale-based linearly combined scaling function* defined at scale $j$ as:

$$\varphi_j(t) = \phi_H\left(2^{j+1}t\right) + \phi_H\left(2^{j+1}\left(t - 1 + 2^{-(j+1)}\right)\right) \tag{4.5}$$

where $j = 1, 2, 3, \ldots$ and $\phi_H(t)$ is the Haar scaling function. The Haar scaling function $\phi_H(t)$ is given by:

$$\phi_H(t) = \begin{cases} 1 & t \in [0, 1) \\ 0 & t \notin [0, 1) \end{cases} \tag{4.6}$$

The selection of the Haar scaling function $\phi_H(t)$ as a building block to design the scale-based linearly combined scaling function $\varphi(t)$ is based on the following features of $\phi_H(t)$:

- The Haar scaling function $\phi_H(t)$ is the only orthogonal scaling function of compact support.
- The Haar scaling function $\phi_H(t)$ is the basic building block for constructing scaling and wavelet functions.

The new designed scale-based linearly combined scaling function $\varphi(t)$ is an $L$th-order scaling function *if and only if* it satisfies the following three conditions. The new designed scale-based linearly combined scaling function $\varphi(t)$ is an $L$th-order scaling function, *if and only if* it satisfies the following three conditions:

- ***Condition 1: Riesz Basis***

$$0 < A \le \hat{a}_\varphi(\omega) < B < +\infty \tag{4.7}$$

where $A$ and $B$ are the Riesz bounds of $\varphi$. The term $\hat{a}_\varphi(\omega)$ is DTFT of $a[k]$, which is an autocorrelation sequence, and is defined as:

$$\hat{a}_\varphi(\omega) = \sum_{k \in \mathbb{Z}} \left|\hat{\varphi}(\omega + 2\pi k)\right|^2 \tag{4.8}$$

where

$$\hat{\varphi}(\omega) \xleftarrow{\;FT\;} \varphi(t) \tag{4.9}$$

$$\hat{a}_\varphi(\omega) \xleftarrow{\;FT\;} a[k] \tag{4.10}$$

*Verification:*

The Fourier transform of the scale-based linearly combined scaling function $\varphi(t)$ is $\hat{\varphi}(\omega)$, which can be determined using Fourier transform properties and $\hat{\phi}_H(\omega)$ (shown in Figure 4.1 (b)) as:

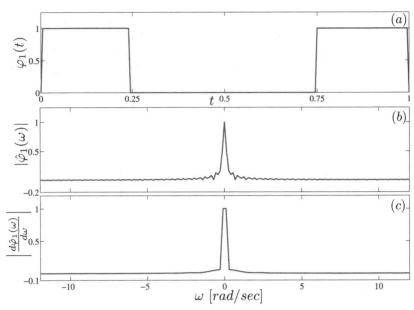

**Figure 4.1**    The scale-based linearly combined scaling function $\varphi(t)$ for $j = 1$: (a) the scaling function $\varphi_1(t)$, (b) the magnitude of its FT $|\hat{\varphi}_1(\omega)|$, and (c) the magnitude of the derivative of $\hat{\varphi}(\omega)\left(\left|\dfrac{d\hat{\varphi}_1(\omega)}{d\omega}\right|\right)$.

$$\hat{\varphi}(\omega) = \frac{1}{2^{(j+1)}}\hat{\phi}_H\left(\frac{\omega}{2^{(j+1)}}\right) + \frac{1}{2^{(j+1)}}\left(e^{-i\omega\left(1-2^{-(j+1)}\right)}\right)\hat{\phi}_H\left(\frac{\omega}{2^{(j+1)}}\right) \tag{4.11}$$

$$\hat{\varphi}(\omega) = \hat{\phi}_H\left(\frac{\omega}{2^{(j+1)}}\right)\left(\frac{1+e^{-i\omega\left(1-2^{-(j+1)}\right)}}{2^{(j+1)}}\right) \tag{4.12}$$

where $i = \sqrt{-1}$. The autocorrelation sequence $\hat{a}_\varphi(\omega)$ can be expressed as:

$$\hat{a}_\varphi(\omega) = \sum_k \left|\hat{\phi}_H\left(\frac{\omega+2\pi k}{2^{(j+1)}}\right)\left(\frac{1+e^{-i(\omega+2\pi k)\left(1-2^{-(j+1)}\right)}}{2^{(j+1)}}\right)\right|^2 \tag{4.13}$$

The autocorrelation sequence $\hat{a}_\varphi(\omega)$ is bounded, since $\hat{\phi}_H(\omega)$ is bounded. Furthermore, the term $\hat{\phi}_H\left(\dfrac{\omega+2\pi k}{4}\right)$ is a decaying function as $\omega \to \infty$. As a result, the Riesz bounds $A$ and $B$ have finite values, which implies that the basis functions generated by $\varphi(t)$ meet the condition for being Riesz basis. These basis functions can be stated as:

$$\{\varphi_{j,k}(t)\} = \{\varphi_1(2^{j-1}t-k), j=1,2\ldots k \in \mathbb{Z}\} \tag{4.14}$$

• **Condition 2:**

$$\varphi(t) = \sqrt{2}\sum_{k \in \mathbb{Z}} h_\varphi[k]\varphi(2t-k) \tag{4.15}$$

where $h_\varphi[k]$ is the refinement filter associated with $\varphi(t)$.

**Verification:**

The refinement filter $h_\varphi[k]$ can be determined by solving the refinement equation from condition 2 that depends on $\phi_H(t)$:

$$\varphi(t) = \sqrt{2} \sum_{k=0}^{L-1} h_\varphi[k] \varphi(2t - k) \tag{4.16}$$

The refinement equation can be rewritten as:

$$\phi(4t) + \phi(4t - 3) = \sqrt{2} \sum_{k=0}^{L-1} h_\varphi[k] \phi(8t - k) + \sqrt{2} \sum_{k=0}^{L-1} h_\varphi[k] \phi(8t - 6 - k) \tag{4.17}$$

The Haar scaling function has ($L = 2$), which makes scaling filter coefficients as:

$$h_\varphi[k] = \left[ \frac{1}{\sqrt{2}} \quad \frac{1}{\sqrt{2}} \right] \tag{4.18}$$

The scaling filter $h_\varphi[k]$ is identical to the Haar scaling filter $(h_\phi)_H[k]$. This result comes due to the linear combination approach in designing $\varphi(t)$.

- **Condition 3:**

$$\left. \frac{d^m \hat\varphi(\omega)}{d\omega^m} \right|_{\omega = 2\pi k} = 0 \tag{4.19}$$

where $k \neq 0$, $m = 0, 1, \dots, L - 1$ and $L$ is the number of vanishing moments of $\varphi(t)$.

**Verrification:**

The derivative of $\hat\varphi(\omega)$ will be:

$$\frac{d\hat\varphi}{d\omega} = \frac{1}{2^{(j+1)}} \left[ \frac{d\hat\phi\left( \frac{\omega}{2^{(j+1)}} \right)}{d\omega} \left( 1 + e^{-i\omega\left(1 - 2^{-(j+1)}\right)} \right) - i\left(1 - 2^{-(j+1)}\right) e^{-i\omega\left(1 - 2^{-(j+1)}\right)} \hat\phi\left( \frac{\omega}{2^{(j+1)}} \right) \right] \tag{4.20}$$

The value of the derivatives of $\hat\varphi(\omega)$ at $\omega = 2\pi k$ are zeros, since $\phi_H(t)$ satisfies the conditions of a scaling function including:

$$\left. \frac{d^m \hat\phi_H}{d\omega^m} \right|_{\omega = 2\pi k} = 0, \quad m = 0, 1, \quad k \neq 0 \tag{4.21}$$

and

$$\left. \left| \hat\phi_H(\omega) \right| \right|_{\omega = 2\pi k} = 0, \quad k \neq 0 \tag{4.22}$$

As a result:

$$\left. \frac{d^m \hat\varphi}{d\omega^m} \right|_{\omega = 2\pi k} = 0, \quad m = 0, 1, \quad k \neq 0 \tag{4.23}$$

The scale-based linearly combined scaling function $\varphi(t)$ is shown in Figure 4.1 along with the magnitude of its FT ($\hat{\varphi}(\omega)$) and $\left|\dfrac{d^m\hat{\varphi}}{d\omega^m}\right|$ (the magnitude of the derivative of its FT).

## 4.2.1  Balancing the Order of the Scale-Based Linearly Combined Scaling Function $\varphi(t)$

The conventional wavelet theory approach suggests that the scale should be set as $j = 0, 1, 2,\dots$. However, evaluating $\varphi(t)$ for $j = 0$ yields:

$$\varphi_0(t) = \phi_H(2t) + \phi_H(2t-1) = \phi_H(t) \tag{4.24}$$

It can be seen from equation (4.24) that $\varphi(t)$ at the scale $j = 0$ is identical to $\phi_H(t)$, in which case the dual scaling function will be the Haar scaling function $\phi_H(t)$ itself. However, for the inverter case the following operation requirements have to be considered:

1. The scale $j = 0$ ($\varphi_0(t) = \phi_H(t)$) is associated with producing a sample group that has one sample. Such sample group does not contribute to the synthesized inverter output, that is, an inverter switching action of 0 duration.

2. The scale $j = 0$ is the scale, where the lowest frequency component of the signal is synthesized, which represent a dc component, which is not present in the inverter output.

3. The inverter is composed of switching elements that require ON and OFF switching times. This requirement can be met by modifying the Haar scaling function as the function responsible for activating the inverter switching element. The Haar scaling function is modified by locating the two samples in each group at:

$$t_{j1} = d + 2^{-(j+1)}$$
$$t_{j2} = d + 1 - 2^{-(j+1)}, \quad d = 1, 2, \dots, (\mathcal{D}_\varphi - 1), j = 1, 2 \dots$$

where $\mathcal{D}_\varphi$ is the number of sample groups created by $\varphi(t)$ over one period of the sampled signal.

4. The inverter output, which represents a synthesized version of the sampled signal (mostly sinusoidal signal(s)), is composed from periodic trains of rectangular pulses, where successive pulses have different widths. This operating requirement indicates that the successive versions of the dual scaling function have to be from different scales at different shifts. This can be realized by creating one version of the dual scaling function required at scale $j$ and shift $k$.

The scaled and translated versions of the scale-based linearly combined scaling function $\varphi(t)$ are shown in Figure 4.2 for several values of the scale $j$ and the integer translations $k$.

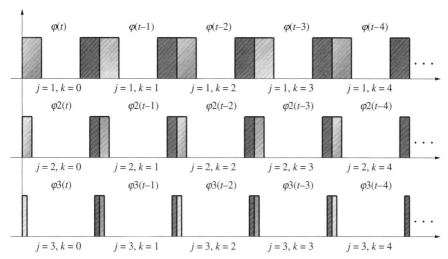

**Figure 4.2**    The scale and translated versions of the scale-based linearly combined scaling function $\varphi(t)$ for several values of the scale $j$ and translation $k$.

The aforementioned operation requirements suggest that the initial value of the scale has to be $j > 0$. Also, these requirements suggest that the dual scaling function has to be a modified form of the Haar scaling function. The modification of the Haar scaling function as the synthesis scaling function can be realized using the locations of the samples created by $\varphi(t)$ at each scale $j$ and shift $k$. On the other hand, the refinement filter associated with the scaling function $\varphi(t)$ is identical to the refinement filter associated with the Haar scaling functions $\phi_H(t)$. This feature is ensured by the linear-combination approach to construct $\varphi(t)$ using $\phi_H(t)$. Also, this feature ensures that both scaling functions $\varphi(t)$ and $\phi_H(t)$ have the same number of vanishing moments. Furthermore, the refinement filter and the number of vanishing moments are consistent due to the scale-based shift of $(\phi_H(t))_{j+1}$. This shift in $\phi_H(t)$ when constructing $\varphi_j(t)$ provides a balance of the order of $\varphi_j(t)$ with respect to the used $\phi_H(t)$. As a consequence, $\varphi(t)$ will have similar properties as $\phi_H(t)$, in particular, the ability to span closed spaces $\{V_j(\varphi)\}$.

The scale-based linearly combined scaling function $\varphi(t)$ features of the scaling filter and number of vanishing moments make it capable of generating basis functions at each scale. These basis functions can be defined as:

$$\left\{\varphi_{j,k}(t)\right\} = \left\{\varphi_1\left(2^{j-1}t - k\right)\right\} j = 1, 2 \ldots k \in \mathbb{Z} \qquad (4.25)$$

The generated basis functions $\{\varphi_{j,k}(t)\}$ span spaces $\{V_j\}$, from a collection of which an MRA can be constructed. However, this constructed MRA will not be dyadic due to the scale-based shift inherent in $\varphi(t)$, and as $j \to \infty$, $\varphi(t) \nrightarrow \delta(t)$.

If a CT periodic signal $x_c(t) \neq 0$, $t \in [t_{1j}, t_{2j}]$ then the inner product $\langle x_c(t),$ $\varphi_{1,k}(t)\rangle$ will have a nonzero value. Limits of the interval of support $[t_1, t_2]$ are defined as:

$$t_{1j} = d + 2^{-(j+1)}$$
$$t_{2j} = d + 1 - 2^{-(j+1)}, \quad d = 1, 2, \ldots, \mathcal{D}_\varphi, j = 1, 2 \ldots \qquad (4.26)$$

where $\mathcal{D}_\varphi$ is the number of sample groups created by $\varphi(t)$ over one period of $S_M(t)$. Note that $\varphi_j(t)$ creates one group of nonuniform samples at each translation $k$. However, for the dc-ac inverter successive *ON* switching pulses have to be of different widths. These pulses aim to reconstruct the CT signal from its nonuniform sample groups. As a consequence, a sample group $d$ is created by $\varphi_1(t)$ scaled to a different scale $j$ that aims to generate different switching pulses over each period of the sampled CT signal $x_c(t)$. The constructed MRA can be generalized as:

$$V_j = \left\{ x_c(t) \neq 0, t \in \left[ t_{1j}, t_{2j} \right]; j = 1, 2, 3, \ldots \right\} \qquad (4.27)$$

provided that $x_c(t)$ is continuous and differentiable over the interval $[t_{1j}, t_{2j}]$. The generalized MRA spaces can be formulated as:

$$V_j(\varphi) = \text{clos}_{L^2} \left\langle \left\{ \varphi_{j,k}(t) \right\} \right\rangle, \quad j = 1, 2, 3, \ldots, k \in \mathbb{Z} \qquad (4.28)$$

The scaling function $\varphi(t)$ is composed of two Haar scaling functions. This indicates that if $\varphi(t)$ is used as a sampling function, two samples will be created at each translation. Each translation depends on both $k$ and the scale-based quantity $(1 - 2^{-(j+1)})$. These translations cause the spacing between samples to be nonuniform. As a consequence, the constructed nondyadic-type MRA can support a nonuniform sampling case.

## 4.2.2   Scale-Based Linearly Combined Wavelet Function $\psi_\varphi(t)$

The definition of the scale-based linearly combined scaling function $\varphi(t)$ represents the basic building block for constructing the nondyadic MRA. However, the complete construction of any MRA is formulated as:

$$V_j = V_{j+1} \oplus W_{j+1} \qquad (4.29)$$

where $\{V_j\}$ are the spaces spanned by the scaling function and where $\{W_j\}$ are the orthogonal complement spaces for $\{V_j\}$. The set of spaces $\{W_j\}$ is spanned by a wavelet function that is associated with the scaling function. In the case of the nondyadic MRA constructed by the scale-based linearly combined scaling function $\varphi(t)$, a wavelet function $\psi_\varphi(t)$ is required to span the spaces $\{W_j\}$ so that:

$$V_j(\varphi) = V_{j+1}(\varphi) \oplus W_{j+1}(\psi_\varphi) \qquad (4.30)$$

The wavelet function $\psi_\varphi(t)$ is defined as the scale-based linearly combined wavelet function and can be determined using the refinement filter associated with $\varphi(t)$ as:

$$\psi_\varphi(t) = \sqrt{2} \sum_{k=0}^{L-1} g_\varphi[k] \varphi(2t - k) \qquad (4.31)$$

where $\varphi(2t)$ is given by:

$$\varphi(2t) = \phi_H(8t) + \phi_H(8t - 6) \qquad (4.32)$$

The vector $g_\varphi[k]$ is related to the refinement filter $h_\varphi[k]$ by the following relation:

$$g_\varphi[k] = (-1)^k \, h_\varphi[L-k] \quad k = 0, 1, \ldots, L-1 \qquad (4.33)$$

Evaluating equation (4.28) with $L = 2$ yields that $g_\varphi[k] = g_\phi[k]$. As a consequence, the wavelet function $\psi_\varphi(t)$ can be expressed as:

$$\psi_\varphi(t) = \sqrt{2} \sum_{k=0}^{L-1} g_\varphi[k] \phi_H(8t - k) + \sqrt{2} \sum_{k=0}^{L-1} g_\varphi[k] \phi_H(8t - 6 - k) \qquad (4.34)$$

Recall the relation between Haar scaling and wavelet functions as:

$$\psi_H(t) = \sqrt{2} \sum_{k=0}^{L-1} g_H[k] \phi_H(2t - k) \qquad (4.35)$$

Using the fact that $g_\varphi[k] = g_H[k]$, the scale-based linearly combined wavelet function can be expressed in terms of the Haar wavelet function $\psi(t)$ as:

$$\left(\psi_\varphi\right)_j(t) = \psi_H\left(2^{j+1} t\right) + \psi_H\left(2^{j+1}\left(t - 1 + 2^{-(j+1)}\right)\right) \qquad (4.36)$$

Figure 4.3 shows scale-based linearly combined wavelet function $\psi_\varphi(t)$ and the magnitude of its Fourier transform $\hat{\psi}_\varphi(\omega)$.

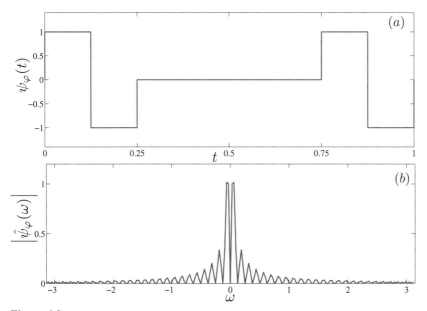

**Figure 4.3**    Scale-based linearly combined wavelet functions: (a) $\varphi_\varphi(t)$ and (b) the magnitude of its FT $|\hat{\psi}_\varphi(\omega)|$.

A signal can be expanded using the generated basis functions by both $\varphi(t)$ and $\psi_\varphi(t)$ as:

$$f(t) = \sum_j \left( \sum_k \langle f(t), \varphi_{j,k}(t) \rangle \tilde{\varphi}_{j,k}(t) + \sum_k \langle f(t), (\psi_\varphi)_{j,k}(t) \rangle (\tilde{\psi}_\varphi)_{j,k}(t) \right) \quad (4.37)$$

where $\tilde{\varphi}_{j,k}(t)$ is a synthesis scaling function and $(\tilde{\psi}_\varphi)_{j,k}(t)$ is the synthesis wavelet function at scale $j$. The inner product term can be written as:

$$\langle f(t), \varphi_{j,k}(t) \rangle = \int_0^{t_{1j}} f(t)(\phi_H)_k(2^{j+1}t)dt + \int_{t_{2j}}^{T_\phi} f(t)(\phi_H)_k(2^{j+1}(t-1+2^{-(j+1)}))dt$$
$$(4.38)$$

where $t_{1j} = d + 2^{-(j+1)}$, and $t_{2j} = d + 1 - 2^{-(j+1)}$, and $T_\phi$ is the interval of support of the Haar scaling function $\phi_H(t)$ at scale $j = 0$. In general, expanding any function as a linear combination of weighted basis functions is a form of series expansion. For the case of the scale-based linearly combined basis functions, each coefficient of such a series is composed of two terms that offer a better and more accurate representation of signals.

### 4.2.3  Construction of Scale-Based Linearly Combined Synthesis Scaling Functions $\tilde{\varphi}(t)$

The linearly combined scaling function $\varphi(t)$ is defined along with its associated refinement filter $h_\varphi[k]$ and wavelet function $\psi_\varphi(t)$. The remaining step toward the complete characterization of the proposed basis functions is to define a linearly combined synthesis scaling and wavelet functions.

The series expansion approach can be used to define a synthesis scaling $\tilde{\varphi}(t)$ and wavelet $\tilde{\psi}_\varphi(t)$ functions. The series expansion of a CT signal $f(t)$ can be written as:

$$\begin{aligned}
f(t) = &\sum_{\substack{j=1,2\ldots \, k\in\mathbb{Z}}} \langle f(t), (\phi_H)_k(2^{j+1}t) \rangle \tilde{\varphi}_{j,k}(t) \\
&+ \sum_{\substack{j=1,2\ldots \, k\in\mathbb{Z}}} \langle f(t), (\phi_H)_k(2^{j+1}(t-1+2^{-(j+1)})) \rangle \tilde{\varphi}_{j,k}(t) \\
&+ \sum_{\substack{j=1,2\ldots \, k\in\mathbb{Z}}} \langle f(t), (\psi_H)_k(2^{j+1}t) \rangle (\tilde{\psi}_\varphi)_{j,k}(t) \\
&+ \sum_{\substack{j=1,2\ldots \, k\in\mathbb{Z}}} \langle f(t), (\psi_H)_k(2^{j+1}(t-1+2^{-(j+1)})) \rangle (\tilde{\psi}_\varphi)_{j,k}(t) \quad (4.39)
\end{aligned}$$

The previous two summations can be expressed in terms of their inner products over the interval $[0, t_{1j}]$ as:

$$\sum_{k\in\mathbb{Z}} \langle f(t), (\phi_H)_k(2^{j+1}t) \rangle \tilde{\varphi}_{j,k}(t) = \sum_{k\in\mathbb{Z}} \left( \int_0^{t_{1j}} f(t)\phi_H(2^{j+1}t-k)dt \right) \tilde{\varphi}(2^jt-k)$$

$$\sum_{k\in\mathbb{Z}} \langle f(t), (\psi_H)_k(2^{j+1}t) \rangle (\tilde{\psi}_\varphi)_{j,k}(t) = \sum_{k\in\mathbb{Z}} \left( \int_0^{t_{1j}} f(t)\psi_H(2^{j+1}t-k)dt \right) \tilde{\psi}_\varphi(2^jt-k)$$
$$(4.40)$$

These two inner products have nonzero values over the interval $[0, t_{1j}]$ that can be interpreted as taking one sample from the signal $f(t)$ over that interval. The other two summations can be expressed in terms of their inner products over the interval $[t_{2j}, T_\phi]$ as:

Let $u_j = 1 + 2^{j+1}$

$$\sum_{k \in \mathbb{Z}} \langle f(t), (\phi_H)_k (2^{j+1} t - u_j) \rangle \tilde{\varphi}_{j,k}(t) = \sum_{k \in \mathbb{Z}} \left( \int_{t_{2j}}^{T_\phi} f(t) \phi_H (2^j t - u_j - k) dt \right) \tilde{\varphi}_{j,k}(t)$$

$$\sum_{k \in \mathbb{Z}} \langle f(t), (\psi_H)_k (2^{j+1} t - u_j) \rangle (\tilde{\psi}_\varphi)_{j,k}(t) = \sum_{k \in \mathbb{Z}} \left( \int_{t_{2j}}^{T_\phi} f(t) \psi_H (2^j t - u_j - k) dt \right) (\tilde{\psi}_\varphi)_{j,k}(t)$$

$$(4.41)$$

These inner products have nonzero values over the interval $[t_{2j}, T_\phi]$, which again can be interpreted as taking another sample of the signal $f(t)$ over the interval $[t_{2j}, 1]$. The summations represent the interpolation process to synthesize $f(t)$ from the samples taken by $\varphi(t)$ using $\tilde{\varphi}_{j,k}(t)$ and $(\tilde{\psi}_\varphi)_{j,k}(t)$ basis functions over the interval $[t_{1j}, t_{2j}]$.

The operation requirements of the inverter can be met by modifying the generation of the Haar basis functions, both analysis and synthesis ones. This modification is established by the scale-based linearly combined scaling function $\varphi(t)$ and its dual synthesis scaling function. The scaling function $\varphi(t)$ still meets the conditions of Riesz basis, generating orthogonal basis functions and refinability. The required modification in the dual synthesis scaling function to satisfy the inverter operational requirements as well as the wavelet theory conditions can be designed based on the following three possibilities, which represent some trade-off between the density of the generated basis functions with accuracy of the reconstruction of the sampled signal $x_c(t)$. These possibilities for designing the dual synthesis scaling function $\tilde{\varphi}_j(t-k)$ are:

- *Possibility 1*: $\tilde{\varphi}_j(t-k)$ has an interval of support as $[0, 1]$, which indicates that $\tilde{\varphi}_j(t-k) = \phi_j(t-k)$. This possibility may not be true due to the structure of $\varphi_j(t-k)$.

- *Possibility 2*: $\tilde{\varphi}_j(t-k) = \varphi_j(t-k)$, which indicates that there will be cross inner products such as:

$$\langle \phi_j(t-k) \tilde{\phi}_j(t - k - (1 - 2^{-(j+1)})) \rangle \neq 0.$$

This possibility also may not be true due to the orthogonality of the used scaling functions $\phi_H(t)$.

- *Possibility 3*: A combination of the previous possibilities, where $\tilde{\varphi}_{j,k}(t)$ has an interval of support related to $\varphi_{j,k}(t)$, can meet orthogonality conditions, the dilation equation, and the structure of $\varphi(t)$. Moreover, $\tilde{\varphi}_{j,k}(t)$ has to have a continuity over its interval of support. This possibility can meet the conditions required for $\tilde{\varphi}_{j,k}(t)$ to be a scaling function.

Using the third possibility, the synthesis scaling function can be defined as:

$$\tilde{\varphi}_j(t) = \phi_H(t) - (\phi_H(2^{j+1} t) + \phi_H(2^{j+1} t - 1 + 2^{j+1}))$$

$$(4.42)$$

The above equation can be expressed in terms of $\varphi_j(t)$ as:

$$\tilde{\varphi}_j(t) = \phi_H(t) - \varphi_j(t), \quad j = 1, 2, \ldots \tag{4.43}$$

The scale-based linearly combined scaling and wavelet functions are capable of spanning orthogonal spaces at each scale $j$ such that:

$$W_j(\psi_\varphi) \perp V_j(\varphi) \tag{4.44}$$

Although the spanned spaces are orthogonal, the constructed MRA is not a dyadic one. This nature of such spaces can construct MRAs for supporting nonuniform sampling forms. The next section presents a nondyadic MRA structure using scale-based linearly combined basis functions.

## 4.3 NONDYADIC MRA STRUCTURE

The defined linearly combined scaling function $\varphi(t)$ can generate sets of basis functions that span successive spaces $\{V_j(\varphi)\}$. Also, the wavelet function $(\psi_\varphi)_j(t)$, associated with $\varphi(t)$, can generate sets of basis functions that span successive spaces $\{W_j(\psi_\varphi)\}$. These spaces satisfy orthogonality and completeness conditions, that is:

$$V_j(\varphi) = V_{j-1}(\varphi) \oplus W_{j-1}(\psi_\varphi) \tag{4.45}$$

The set of spaces $\{V_j(\varphi)\}$, composed of orthogonal spaces spanned by scaling and wavelet basis functions, has a nested structure such that:

$$\ldots \subset V_{-1} \subset V_0 \subset V_1 \subset V_2 \ldots \subset V_m \subset L^2(\mathbb{R}) \tag{4.46}$$

The collection of such spanned spaces constructs a stable MRA. Note that both $\varphi(t)$ and $\psi_\varphi(t)$ have dilations as the level $j$ changes. Also, both $\varphi(t)$ and $\psi_\varphi(t)$ have translations that are created by the change in $k$ as well as $(1 - 2^{-(j+1)})$. These types of dilations and translations create a unique nondyadic MRA. Such an MRA can be employed to support nonuniform sampling forms. The next subsection provides the nonuniform recurrent sampling form that is supported by the developed nondyadic MRA.

### 4.3.1 MRA for Nonuniform Recurrent Sampling

The scale-based linearly combined scaling function $\varphi(t)$ creates two samples over its interval of support. Moreover, the spacing between these samples depends on the scale $j$. A CT signal $x_c(t)$ can be reconstructed from its nonuniform recurrent samples $x_c(t_p)$ using the Lagrange general interpolation formula as:

$$x_c(t) = \sum_{n=-\infty}^{\infty} \sum_{p=1}^{2} x(t_p) \frac{G(t)}{G'(t_p)(t - t_p)} \tag{4.47}$$

An interpolation function $\lambda(t)$ can be defined as:

$$\lambda_j(t) = \sum_{p=1}^{2} x(t_{pj}) \frac{G(t)}{G'(t_{pj})(t - t_{pj})} \tag{4.48}$$

The two samples created by $\varphi(t)$ at each scale $j$ can be considered as one group of nonuniform samples that can be described by:

$$\{t_{pj}\} = \{t_{1j} = d + 2^{-(j+1)}, t_{2j} = d + 1 - 2^{-(j+1)}\}, \quad p = 1, 2, \quad d = 1, 2 \ldots, \mathcal{D}_\varphi \qquad (4.49)$$

The set of sampling instants $\{t_{pj}\}$ for the sample group $d$ that is related to the scale $j$ can be used to evaluate the function $G(t)$ at the scale $j$, $(j = 1, 2, 3,\ldots, \infty)$ as:

$$G_j(t) = t\left(1 - \frac{t}{2^{-j-1}}\right)\left(1 - \frac{t}{1 - 2^{-j-1}}\right) \qquad (4.50)$$

It should be noted that the value of $t_0$ is considered zero to indicate the beginning of each cycle of the sampled CT signal $x_c(t)$. The function $G(t)$ can be simplified to:

$$G_j(t) = t - \frac{2^{2j+2}t^2}{2^{j+1} - 1} + \frac{2^{2j+2}t^3}{2^{j+1} - 1} \qquad (4.51)$$

Evaluating the derivative of $G(t)$ at $t_1$ and $t_2$ gives the following values:

$$G_j'(t_{1j}) = \frac{1}{2^{j+1} - 1} - 1 \qquad (4.52)$$

$$G_j'(t_{2j}) = 2^{j+1} - 2 \qquad (4.53)$$

The quantities $G_j(t)$, $G_j'(t_1)$, and $G_j'(t_2)$ can be employed to evaluate the interpolation function $(\lambda_j(t))$ for $j = 1$:

$$\lambda_{j=1}(t) = -x(t_{11})\frac{3t - 16t^2 + 16t^3}{2\left(t - \frac{1}{4}\right)} + x(t_{12})\frac{3t - 16t^2 + 16t^3}{6\left(t - \frac{3}{4}\right)} \qquad (4.54)$$

For $j = 2$, the interpolation function $\lambda_j(t)$ is:

$$\lambda_{j=2}(t) = -x(t_{21})\frac{t - 64t^2 + 64t^3}{\frac{6}{7}\left(t - \frac{9}{8}\right)} + x(t_{22})\frac{t - 64t^2 + 64t^3}{6\left(t - \frac{15}{8}\right)} \qquad (4.55)$$

Figure 4.4 shows the two interpolation functions $\lambda_{j=1}(t)$ and $\lambda_{j=2}(t)$.

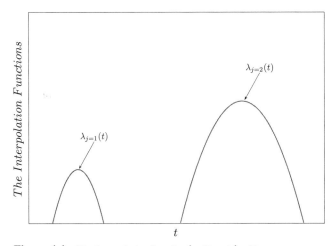

**Figure 4.4**   The interpolation function $\lambda_{j=1}(t)$ and $\lambda_{j=2}(t)$.

The two examples of the interpolation function $\lambda_j(t)$ provided in equations (4.110) and (4.111) suggest that at each scale $j$ and translation $k$, there is one group of samples created. Also, it indicates that a set of interpolation functions $\{\lambda_j(t)\}_{j=1,2...}$ will have a set of intervals of support such that $t \in [t_{1j}, t_{2j}]$, where $t_j$ indicates that the interpolation is done over the sample group $d$ that is created by $\varphi(t)$ at scale $j$. Moreover, each interpolation function is continuous and differentiable over its interval of support.

In general, inverters are composed of switching elements that are operated either fully ON or fully OFF. As a result, inverter outputs are trains of rectangular pulses with variable widths and locations, which represent a synthesized version of the sinusoidal reference-modulating signal $S_M(t)$. The sampling-based inverter model developed in Chapter 3 defines these rectangular pulses as stages in an interpolation process, which is consistent with the developed nondyadic MRA. Each interpolation function $\lambda_j(t)$ defines an ON switching pulse. This ON pulse will be translated to a rectangular pulse by an activated switching element(s) appearing on inverter output terminals. Such output due to $\lambda_j(t)$ comes consistent with defined scale-based linearly combined synthesis scaling function $\tilde{\varphi}(t)$. Figure 4.5 shows the interpolating function $\lambda_j(t)$ for $j = 2$ and the corresponding synthesis scale-based linearly combined scaling function $\tilde{\varphi}(t)$.

Figure 4.6 shows both the scaling functions $\varphi(t)$ and $\tilde{\varphi}(t)$ for $j = 1$, 2 and $k = 0$, 1. Two samples are created over each interval of support for $\varphi(t)$. Furthermore, the spacing between the samples in each sample group increases as the scale $j$ increases. Note that the change in the scale $j$ affects the translation of $\varphi_j(t)$, but does not allow successive groups of samples to overlap. The MRA associated with this form of sampling can be created by the defined scale-based linearly combined scaling function $(\varphi_j(t))$. The reconstruction of a CT signal $x_c(t)$ from its nonuniform

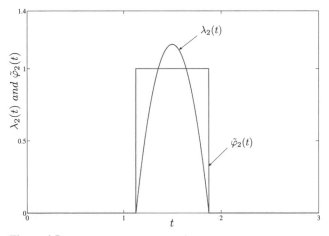

**Figure 4.5**    The interpolation function $\lambda_j(t)$ for $j = 2$ and the corresponding synthesis scaling function $\tilde{\varphi}_2(t)$.

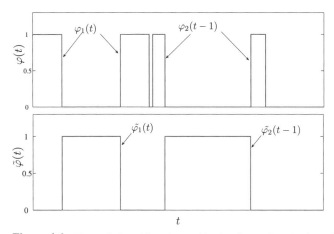

**Figure 4.6**    The scale-based linearly combined scaling and synthesis scaling functions $\varphi(t)$ and $\tilde{\varphi}(t)$ for $j = 1, 2$ and $k = 0, 1$.

recurrent samples the scale-based linearly combined synthesis scaling function using $\tilde{\varphi}(t)$ can be expressed as:

$$x_c(t) = \sum_j \sum_{k \in \mathbb{Z}} \langle x_c(t), \varphi_j(t-k) \rangle \tilde{\varphi}_j(t-k) \tag{4.56}$$

The defined scale-based linearly combined scaling functions $\varphi(t)$ and $\tilde{\varphi}(t)$ as well as their corresponding wavelet function $\psi_\varphi(t)$ are capable of spanning spaces that define a stable nondyadic MRA to support a nonuniform recurrent sampling. Note that the structure of $\varphi(t)$ using $\phi_H(t)$ guarantees creating two samples over its interval of support at each scale $j$. This consideration is important for meeting the requirements of the sampling-based inverter model developed in Chapter 3. However, in other applications, such scaling and wavelet functions may be constructed to create more than two samples using other scaling functions than the Haar scaling function.

## 4.4 SCALE-BASED LINEARLY COMBINED SCALING FUNCTIONS FOR THREE-PHASE INVERTERS

Inverters can be designed as single-phase ($1\phi$) or three-phase ($3\phi$) configurations. Three-phase inverters are very popular in a wide range of industrial applications such as power systems, ac motor drives, renewable energy utilization, and power supplies. The developed scale-based linearly combined wavelet basis-function-based nondyadic-type MRAs can be extended to generate switching pulses for operating $3\phi$ inverters. The basis for the desired MRAs will depend on the nonuniform recurrent sampling model of $3\phi$ inverters developed in Chapter 3, in particular, sampling shifted CT signals.

The extension of the nonuniform recurrent sampling model for $3\phi$ inverters is based on considering the $3\phi$ inverter as three independent $1\phi$ inverters. This consideration is justified by the principle of operating $3\phi$ inverters, where the switching pulses are generated using reference-modulating signals representing $3\phi$ line quantities. Such foundation of modeling $3\phi$ inverters can be incorporated for developing nondyadic-type MRAs to generate switching pulses for operating a $3\phi$ inverter.

The basis of the wavelet modulation technique lies in the definition of the scale-based linearly combined scaling function $\varphi_1(t)$, which is used to sample the reference-modulating signal in a nonuniform recurrent manner. In a $3\phi$ sixpulse inverter, three reference-modulating signals are used to generate the required switching pulses. As a consequence, three scaling functions are needed to sample these three signals. Note that these three reference-modulating signals have the same frequency and magnitude with a phase shift of $\dfrac{2\pi}{3}$ from each other. The three reference-modulating signals, each one of them related to one phase of the desired $3\phi$ output, can be defined as:

$$S_{Ma}(t) = \sin(\omega_m t) \tag{4.57}$$

$$S_{Mb}(t) = \sin\left(\omega_m t - \frac{2\pi}{3}\right) \tag{4.58}$$

$$S_{Mc}(t) = \sin\left(\omega_m t + \frac{2\pi}{3}\right) \tag{4.59}$$

The nonuniform recurrent sampling of these three CT reference-modulating signals can be achieved using one sampling function for each CT signal. As a consequence, three scaling functions with a phase shift of $\dfrac{2\pi}{3}$ from each other are required to sample the three reference-modulating signals in a nonuniform recurrent manner. These three scale-based linearly combined scaling functions can be defined as:

$$\big(\varphi_a(t)\big)_j = \phi_H\big(2^{j+1}t\big) + \phi_H\big(2^{j+1}(t-u)\big) \tag{4.60}$$

$$\big(\varphi_b(t)\big)_j = \phi_H\big(2^{j+1}t - z_b\big) + \phi_H\big(2^{j+1}(t-u) - z_b\big) \tag{4.61}$$

$$\big(\varphi_c(t)\big)_j = \phi_H\big(2^{j+1}t - z_c\big) + \phi_H\big(2^{j+1}(t-u) - z_c\big) \tag{4.62}$$

where $u = 1 - 2^{-(j+1)}$, $j = 1, 2,...$ and the shifts $z_b$ and $z_c$ are given by:

$$z_b = \frac{2^{2-j}}{3} \tag{4.63}$$

$$z_c = \frac{2^{3-j}}{3} \tag{4.64}$$

Figure 4.7 shows the three scaling functions $\varphi_a(t)$, $\varphi_b(t)$, and $\varphi_c(t)$ for $j = 1, 2$. The three scale-based linearly combined scaling functions $(\varphi_a(t))$, $(\varphi_b(t))$, and $(\varphi_c(t))$ are capable of creating sets of nonuniform recurrent sample groups for the three reference-modulation signals $S_{Ma}(t)$, $S_{Mb}(t)$, and $S_{Mc}(t)$. Note that the three scale-based

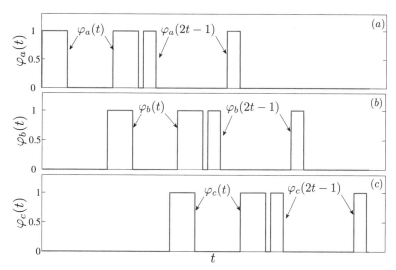

**Figure 4.7**    The three scale-based linearly combined scaling functions for $j = 1, 2$: (a) $\varphi_a(t)$, (b) $\varphi_b(t)$, and (c) $\varphi_c(t)$.

linearly combined scaling functions are required to meet the conditions for constructing an MRA using dilated and translated versions of each scaling function. This condition is necessary to ensure the reconstruction of the CT signals using dilated and translated versions of the dual synthesis scaling functions.

Reconstructing the three CT reference-modulation signals $S_{Ma}(t)$, $S_{Mb}(t)$, and $S_{Mc}(t)$ is carried out by the synthesis scaling functions, where each one of them is the dual of one scale-based linearly combined scaling function. These scale-based linearly combined synthesis scaling functions can be defined as:

$$\left(\tilde{\varphi}_a(t)\right)_j = \phi_H(t) - \left(\varphi_a(t)\right)_j \tag{4.65}$$

$$\left(\tilde{\varphi}_b(t)\right)_j = \phi_H(t - z_b) - \left(\varphi_b(t)\right)_j \tag{4.66}$$

$$\left(\tilde{\varphi}_c(t)\right)_j = \phi_H(t - z_c) - \left(\varphi_c(t)\right)_j \tag{4.67}$$

where $j = 1, 2, \ldots$. Figure 4.8 shows the three synthesis scaling functions $\tilde{\varphi}_a(t)$, $\tilde{\varphi}_b(t)$, and $\tilde{\varphi}_c(t)$ for $j = 1, 2$. These synthesis scaling functions are used to activate the switching elements of the $3\phi$ inverter. Each of the three scaling functions $(\varphi_a(t))_j$, $(\varphi_b(t))_j$, and $(\varphi_c(t))_j$ is composed of two Haar scaling functions. This feature indicates that each scaling function creates two samples at each dilation $j$ and translation $k$. On the other hand, changing the scale $j$ will cause both dilations and translations of each scaling function. Furthermore, each translation is dependent on both the translation integer $k$ and the scale-dependent quantity $(1 - 2^{-(j+1)})$. These translations cause the spacing between samples to be nonuniform. This feature guarantees that synthesis scaling functions will have different locations with different intervals of support.

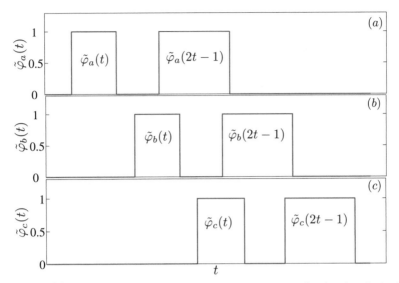

**Figure 4.8**    The three linearly combined scale-based synthesis scaling functions for $j = 1, 2$: (a) $\tilde{\varphi}_a(t)$, (b) $\tilde{\varphi}_b(t)$, and (c) $\tilde{\varphi}_c(t)$.

Using the three developed scale-based linearly combined scaling functions $(\varphi_a(t))_j$, $(\varphi_b(t))_j$, and $(\varphi_c(t))_j$ and their dual synthesis scaling functions $(\tilde{\varphi}_a(t))_j$, $(\tilde{\varphi}_b(t))_j$, and $(\tilde{\varphi}_c(t))_j$, the three CT reference-modulating signals $S_{Ma}(t)$, $S_{Mb}(t)$, and $S_{Mc}(t)$ can be expanded as:

$$S_{Ma}(t) = \sum_j \sum_{k \in \mathbb{Z}} \left\langle S_{Ma}(t), (\varphi_a(t-k))_j \right\rangle (\tilde{\varphi}_a(t-k))_j \qquad (4.68)$$

$$S_{Mb}(t) = \sum_j \sum_{k \in \mathbb{Z}} \left\langle S_{Mb}(t), (\varphi_b(t-k))_j \right\rangle (\tilde{\varphi}_b(t-k))_j \qquad (4.69)$$

$$S_{Mc}(t) = \sum_j \sum_{k \in \mathbb{Z}} \left\langle S_{Mc}(t), (\varphi_c(t-k))_j \right\rangle (\tilde{\varphi}_c(t-k))_j \qquad (4.70)$$

The inner products in equations (4.65), (4.66), and (4.67) can be written as:

$$\left\langle S_{Ma}(t), (\varphi_a(t))_j \right\rangle = \int_0^{t_{ad1}} S_{Ma}(t)(\phi_H)_{j+1}(t)dt + \int_{t_{ad2}}^{T_\phi} S_{Ma}(t)\phi_H(l)dt \qquad (4.71)$$

$$\left\langle S_{Mb}(t), (\varphi_b(t))_j \right\rangle = \int_0^{t_{bd1}} S_{Mb}(t)(\phi_H)_{j+1}(t-z_b)dt + \int_{t_{bd2}}^{T_\phi+z_b} S_{Mb}(t)\phi_H(l-z_b)dt \qquad (4.72)$$

$$\left\langle S_{Mc}(t), (\varphi_c(t))_j \right\rangle = \int_0^{t_{cd1}} S_{Mc}(t)(\phi_H)_{j+1}(t-z_c)dt + \int_{t_{cd2}}^{T_\phi+z_c} S_{Mc}(t)\phi_H(l-z_c)dt \qquad (4.73)$$

where $\phi_H(l) = \phi_H(2^{j+1}(t - 1 + 2^{-(j+1)}))$. The aforementioned inner products ensure the creation of two samples over the interval of support of each scaling function at each scale $j$ and translation $k$. It should be noted that scale $j$ has to start from $j_0 = 1$ in order to avoid uniform sampling that takes place for $j = 0$.

## 4.5  SUMMARY

There are different types of wavelet basis functions that are capable of spanning spaces to construct MRAs. In general, these MRAs are based on dyadic structure that can only support uniform sampling. Wavelet basis functions can be designed using the dilation matrix-based method, convolution of dual basis functions, fractional *B-spline* basis functions, and the autocorrelation sequence-based method. These methods of designing wavelet basis functions guarantee different combinations of scaling and wavelet spaces within dyadic MRA structures. The need for nondyadic MRAs in applications that include modeling inverters has motivated the development of new types of scaling and wavelet basis functions to construct such types of MRA.

The new scale-based linearly combined basis functions are constructed using a new design method, which is based on a linear combination of dilated scaling and wavelet functions. This design method has produced a new type of wavelet basis functions that has been verified to support a nonuniform recurrent sampling-reconstruction process. The newly designed scale-based linearly combined basis functions are used to develop a modulation technique to operate inverters. Furthermore, this chapter has presented the development and construction of an extended nondyadic-type MRA that is responsible for sampling and reconstructing three CT reference-modulating signals, which are associated with the three phases on the $3\phi$ inverter outputs. This approach for operating the $3\phi$ inverter has been entirely based on the development of the sampling-based model for $3\phi$ inverters.

# Chapter 5

# Single-Phase Wavelet Modulated Inverters

## 5.1 INTRODUCTION

The fundamental function of an inverter is to convert dc voltages into ac ones. This function is basically achieved through activating inverter switching elements in a sequential manner that guarantees periodic changes of the output voltage polarity. Several techniques have been developed and tested to operate inverters to produce output voltages as close to sinusoidal waveforms as possible. One of the major concerns of operating an inverter is the presence of harmonic components on the inverter output side. In general, achieving inverter outputs with reduced harmonics has been considered as a tradeoff between the complexity of the switching scheme and the efficiency of the inverter itself. A possible approach for reducing such harmonic components can be based on developing a modulation technique that is correlated with an accurate modeling of the inverter.

In this chapter, a new modulation technique for operating single-phase ($1\phi$) inverters is implemented and tested. This modulation technique is called the wavelet modulation that is based on the nondyadic wavelet-based multiresolution analysis (MRA) constructed using the scale-based linearly combined basis functions, which are described in Chapter 4. Furthermore, the wavelet modulation technique is correlated with the inverter sampling-based model developed in Chapter 3.

## 5.2 IMPLEMENTING THE WAVELET MODULATION TECHNIQUE

The basic correlation between the wavelet modulation technique and the inverter sampling-based model is created through the instantaneous responses of the inverter switching elements. The responses of the switching elements represent the synthesis part of the nonuniform recurrent sampling-reconstruction process, which characterizes the overall inverter operation. On the other hand, inverter switching

*An Introduction to Wavelet Modulated Inverters*, by S.A. Saleh and M. Azizur Rahman
Copyright © 2011 Institute of Electrical and Electronics Engineers

elements need time for ON and OFF switching times in order to change their status from ON to OFF or OFF to ON. Such practical constraints of operating an inverter are inherently provided by the scale-based linearly combined synthesis functions $\tilde{\varphi}_{j,k}(t)$, which are defined over time intervals of sample groups. These sample groups are separated by small time intervals due to locations of samples that can provide the switching elements with the needed switching times. The other constraint for operating an inverter is the energy in the reconstructed signal, which in this case is the inverter output. In general, the wavelet modulation technique has to be capable of concentrating the energy of the reconstructed signal in the fundamental frequency component. This constraint ensures that minimum energy be distributed in undesired frequency bands. If these constraints are taken into account when operating an inverter, output voltage and current can have very low harmonic components.

The basis of the wavelet modulation technique lies in the construction of a nondyadic MRA, in particular, sampling the reference-modulating signal in a non-uniform recurrent manner, and then reconstructing it through the inverter switching actions. The implementation of the wavelet modulation technique can be divided into two parts:

- Implementing a nonuniform recurrent sampling with a period of $T_m$, which is the period of the reference-modulating signal $S_M(t)$. These samples are created by dilated and shifted versions of the scale-based linearly combined scaling function $\varphi_1(t)$ and arranged in groups of two samples each.
- Generating switching pulses that are dilated and shifted versions of the synthesis scaling function $\tilde{\varphi}_1(t)$.

The developed scale-based linearly combined scaling function $\varphi_{j,k}(t)$ creates a group of two samples at each dilation (change in scale $j$) and shift (change in $k$). For each cycle of $S_M(t)$, a finite number of sample groups $\mathcal{D}_\varphi$ is created. The time interval of each group can be defined as:

$$t \in \left[ t_{1j}, t_{2j} \right] \tag{5.1}$$

where $t_{1j}$ and $t_{2j}$ are the time locations of the first and second samples of the sample group $d$ created by $\varphi_j(t)$, respectively. Also, the time interval of each sample group represents the interval of support for the scaling function $\varphi_1(t)$ at scale $j$ and shift $k$. The length of the time interval of the sample group changes as the scale $j$ changes. This relation between the scale $j$ and the time interval of each sample group can be stated as:

$$t_{1j} = d + 2^{-(j+1)}$$
$$t_{2j} = d + 1 - 2^{-(j+1)}, d = 1, 2, \ldots, \mathcal{D}_\varphi, j = 1, 2, \ldots \tag{5.2}$$

Note that the scale $j$ has to start with $j_0 = 1$ in order to avoid uniform sampling that takes place for $j = 0$.

The reconstruction of the reference-modulating signal $S_M(t)$ from its nonuniform recurrent samples is carried out using dilated and shifted versions of the synthesis

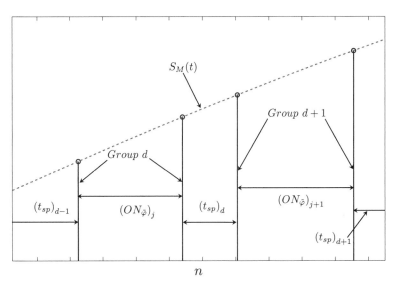

**Figure 5.1**    Two groups ($d$ and $d + 1$) of nonuniform recurrent samples and the location of the time interval $(t_{sp})_d$ that separates them. Also, the durations on the associated ON switching pulses $(ON_{\tilde{\varphi}})_j$ and $(ON_{\tilde{\varphi}})_{j+1}$.

scaling function $\tilde{\varphi}(t)$. Each dilated and shifted version of the synthesis scaling function $\tilde{\varphi}(2^j t - k)$ has an interval of support that is given by:

$$t \in \left[t_{1j}, t_{2j}\right] \tag{5.3}$$

Both scaling functions $\varphi_{j,k}(t)$ and $\tilde{\varphi}_{j,k}(t)$ at scale $j$ and shift $k$ have identical intervals of support. This property is consistent with properties of the Haar scaling function $\phi_H(t)$ that is used to construct $\varphi(t)$.

The synthesis scaling function $\tilde{\varphi}_{j,k}(t)$ at scale $j$ and shift $k$ creates an ON switching pulse with a duration $ON_{\tilde{\varphi}}$ that can be defined as:

$$ON_{\tilde{\varphi}} = t_{2j} - t_{1j} \tag{5.4}$$

Note that the location $d$ is taken with respect to an arbitrary reference. In general, inverter switching elements are activated by successive sets of synthesis scaling functions $\{\tilde{\varphi}_{j,k}(t)\}$ that have their adjacent intervals of support placed such that:

$$t_{d1} = t_{sp} + t_{(d-1)2} \tag{5.5}$$

where $t_{sp}$ is a time interval that separates the intervals of support for the sample groups $d$ and $d - 1$. The set of these time intervals $\{t_{sp}\}_d$ can provide switching elements with the needed ON and OFF switching times. Figure 5.1 shows two sample groups and their time intervals and the location of the time interval $(t_{sp})_d$ that separates them.

The wavelet modulation technique can be realized through a procedure with the following steps:

- **Step 1**: Set the scale $j$ to $j_0 = 1$ and initialize $\gamma$ to $\gamma_0 = 1$ and the index of sample groups $d$ to $d_0 = 0$.
- **Step 2**: Create one sample at $t_{d1} = d + 2^{-(j+1)}$, and another sample at $t_{d2} = d + 1 - 2^{-(j+1)}$.
- **Step 3**: Generate an ON switching pulse over the interval of support for the sample group $d$.
- **Step 4**: Evaluate the derivative $S'_M(t)\big|_{t_{d2}}$ as:

$$S'_M(t_{d2}) = \cos(\omega_m t_{d2}) \tag{5.6}$$

  - If $S'_M(t_{d2}) > 0$, the scale $j$ is increased by 1 for the next samples group.
  - If $S'_M(t_{d2}) < 0$, the scale $j$ is decreased by 1 for the next samples group.
- **Step 5**: Calculate $\gamma_d$ as:

$$\gamma_d = \gamma_{d-1} + \frac{S'_M(t_{d2})}{|S'_M(t_{d2})|} \sum_{m=1}^{j} m2^m (t_{m2} - t_{m1}) \tag{5.7}$$

where:

$$t_{m1} = 2^{-(m+1)}$$

$$t_{m2} = 1 - 2^{-(m+1)}$$

- **Step 6**: If $t \geq T_m$, set $j$ to $j_0$ and $d$ to $d_0$. Otherwise, increase the index $d$ by 1.
- **Step 7**: If $\gamma_d \leq 1$ then $Q_1(t) = Q_3(t) = 1$ and
  $Q_2(t) = Q_4(t) = 0$: The dead-beat condition.
- **Step 8**: Go to Step 2.

The evaluation of $S'_M(t)\big|_{t_{2j}}$ is needed to ensure that both the scale $j$ and number of sample groups $\mathcal{D}_\varphi$ over one cycle of $S_M(t)$ have finite maximum values. The aforementioned procedure to implement the wavelet modulation technique is converted into a flowchart that is shown in Figure 5.2.

## 5.3  SIMULATED PERFORMANCE OF A WAVELET MODULATED INVERTER

The previous section presented a procedure for implementing the wavelet modulation technique for operating a single-phase inverter through processing (sampling and reconstructing) the CT signal $S_M(t)$. This procedure can be converted into an executable program for performance testing purposes. Such an executable program carries out sampling $S_M(t)$ using dilated and shifted versions of the scaling function $\varphi_{j,k}(t)$. Also, it can generate switching pulses using dilated and shifted versions of the scaling function $\tilde{\varphi}_{j,k}(t)$.

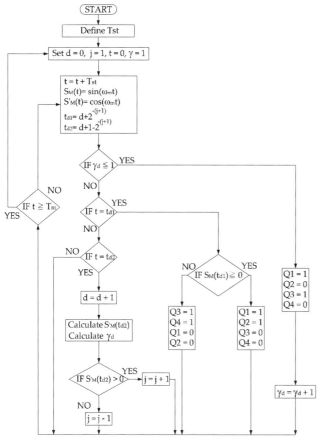

**Figure 5.2**    A flowchart for an algorithm to implement the wavelet modulation technique. Time instants $t_{d1}$ and $t_{d2}$ are defined in equation (5.2). The reference-modulating signal $S_M(t)$ is given by $S_M(t) = \sin(\omega_m t)$.

In order to simulate the performance of a single-phase ($1\phi$) voltage-source (VS) four-pulse (H-bridge) inverter operated by the wavelet modulation technique, the MATLAB/SIMULINK is used. The tested loads are as follows:

- Linear load: a resistive load with a resistance $R_L$ as $R_L = 15\,\Omega$
- Linear load: an $R - L$ load with an impedance $Z_L$ as $Z_L = 15 + j12.88\,\Omega$
- Dynamic load: a $1\phi$, 0.75-hp, 60 Hz, 120 V, 2400 RPM universal motor
- Nonlinear load: a $1\phi$ full-wave diode rectifier with a resistive load of $25\,\Omega$ (nonlinear load)

The quality of inverter output voltage and load current is usually expressed in terms of the total harmonic distortion factor (THD). This factor is defined as:

$$THD_A = \frac{A_h}{A_1} \tag{5.8}$$

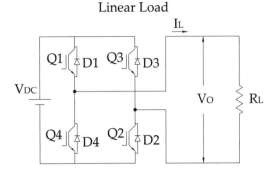

Linear Load

Figure 5.3 Schematic diagram of a $1\phi$ $H$-bridge IGBT inverter and the resistive load.

where $A_1$ is the *RMS* value of the fundamental component of the inverter output voltage or load current. $A_h$ is the summation of all other harmonic components *RMS* values that is given by:

$$A_h = \sqrt{\left(A_2^2 + A_3^2 + \ldots + A_n^2\right)} \qquad (5.9)$$

where $h = 2, 3, \ldots, n$ indexes the harmonic components present in $A$.

### The Resistive Load (Linear Load)

The procedure of implementing the wavelet modulation technique in Section 5.2 is realized using MATLAB/Simulink to simulate the performance of a $1\phi$ VS $H$-bridge inverter. The simulated inverter is activated by switching pulses generated by the wavelet modulation technique (called wavelet modulated (WM) inverter), and has an input dc voltage of 50 V to supply the resistive load. This simulated test is meant to investigate the open-loop steady-state performance of the WM inverter when supplying a linear load. Figure 5.3 shows the schematic diagram of the $1\phi$ voltage source (VS) insulated gate bipolar transition (IGBT) $H$-bridge inverter and the resistive load. Figure 5.4 shows the resistive load voltage $V_L(t)$ and current $I_L(t)$ along with their spectra obtained using a Matlab built-in FFT function to determine their THD$_V$ and THD$_I$ factors.

The spectra in Figure 5.4 show that significant amounts of energy are concentrated in the voltage and current fundamental components, while almost negligible energy is allocated in harmonic components. The features observed in the voltage and current indicate the ability of the WM inverter to transfer higher power from the supply to the load.

### The R – L Load (Linear Load)

This simulated performance of the WM inverter is carried out by connecting an inductive $R - L$ of an impedance of $Z_L = 15 + j12.88\,\Omega$ on the output of the inverter as shown in Figure 5.5. The $R - L$ load simulated test is meant to investigate the open-loop steady-state performance of the WM inverter when supplying a linear

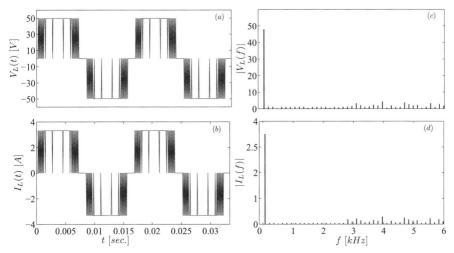

**Figure 5.4** The $R - L$ load: (a) the load voltage $V_L(t)$, (b) the spectrum of $V_L(t)$, (c) the $R - L$ load current $I_L(t)$, and (d) the spectrum of $I_L(t)$. $|V_1| = 48.24\,V$, $THD_V = 17.68\%$, $|I_1| = 3.22\,A$ and $THD_I = 17.68\%$.

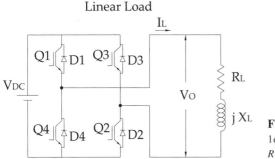

Linear Load

**Figure 5.5** Schematic diagram of a $1\phi$ H-bridge IGBT inverter and the $R - L$ load.

load. Figure 5.6 shows the $R - L$ load voltage $V_L(t)$ and current $I_L(t)$ along with their spectra.

The spectra in Figure 5.6 confirm the ability of the WM inverter to concentrate significant amounts of energy is concentrated in the voltage and current fundamental components. Furthermore, the spectra of Figure 5.6 show that almost negligible energy is allocated in harmonic components. The features observed in the voltage and current indicate the ability of the WM inverter to transfer higher power from the supply to the load.

### The Single-Phase Universal Motor (Dynamic Load)

This performance simulation test is carried out to investigate the open-loop steady-state performance of the WM inverter when supplying a dynamic load. The simu-

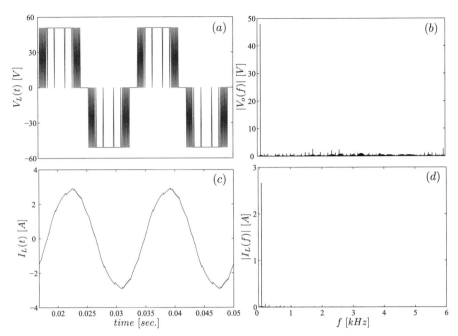

**Figure 5.6**   The $R - L$ load: (a) the load voltage $V_L(t)$, (b) the spectrum of $V_L(t)$, (c) the $R - L$ load current $I_L(t)$, and (d) the spectrum of $I_L(t)$. $|V_1| = 48.24\,V$, $\text{THD}_V = 17.68\%$, $|I_1| = 2.37\,A$, and $\text{THD}_I = 2.13\%$.

lated inverter is activated by switching pulses generated by the WM technique, and has an input dc voltage of 120 V to supply the dynamic load. The used dynamic load is a $1\phi$, 120 V, 60 Hz, 0.75 hp, 2400 RPM universal motor as shown in Figure 5.7. The inverter output voltage $V_M(t)$ and motor current $I_M(t)$ along with their spectra are shown in Figure 5.8.

The spectra in Figure 5.8 show consistency for significant amounts of energy being concentrated in the voltage and current fundamental components, while negligible energy is allocated in harmonic components. The features observed in the voltage and current produce low motor current as a result of rated voltage applied on the motor terminal.

### The Diode Rectifier (Nonlinear Load)

The nonlinear load performance simulated test is carried out to investigate the open-loop steady-state performance of the WM inverter when supplying a nonlinear load. The simulated inverter is activated by switching pulses generated by the wavelet modulation technique, and has an input dc voltage of 50 V to supply the nonlinear load. The nonlinear load used for this test is a $1\phi$ full-wave diode rectifier that has

Dynamic Load

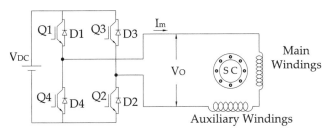

**Figure 5.7**   Schematic diagram of a 1$\phi$ H-bridge IGBT inverter and the 1$\phi$, 120 V, 60 Hz, 0.75 hp, 2400 RPM universal motor.

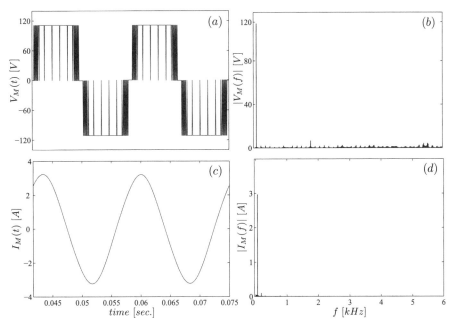

**Figure 5.8**   The 1$\phi$ universal motor: (a) the motor voltage $V_M(t)$, (b) the spectrum of $V_M(t)$, (c) the motor current $I_M(t)$, and (d) the spectrum of $I_M(t)$. $|V_1| = 108.38$ V, THD$_V$ = 17.68%, $|I_1| = 3.1$ A, and THD$_I$ = 1.69%.

a resistive load of $R = 25\,\Omega$ as shown in Figure 5.9. Figure 5.10 shows the nonlinear load voltage $V_L(t)$ and current $I_L(t)$ along with their spectra.

The simulated test results for the nonlinear load shown in Figure 5.10 confirm the features of the load voltage and current observed for the other simulated load types.

### Nonlinear Load

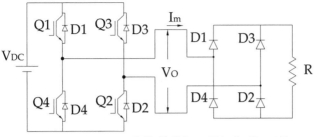

### A Full-Wave Diode Rectifier

**Figure 5.9**    Schematic diagram of a $1\phi$ $H$-bridge IGBT inverter and the $1\phi$ full-wave diode rectifier.

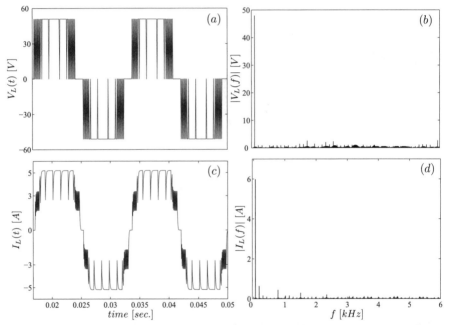

**Figure 5.10**    The nonlinear load: (a) the load voltage $V_L(t)$, (b) the spectrum of $V_L(t)$, (c) the load current $I_L(t)$, and (d) the spectrum of $I_L(t)$. $|V_1| = 46.35\,V$, $\text{THD}_V = 14.68\%$, $|I_1| = 6.23\,A$, and $\text{THD}_I = 26.43\%$.

The performance simulation tests have demonstrated quite encouraging results in terms of harmonic contents of inverter output voltages and currents. The total harmonic distortion factors ($\text{THD}_V$ and $\text{THD}_I$) evaluated for several tests show significant reduction of the energy distributed in the harmonic frequencies that indicates producing high-quality outputs and high power transfer from the supply side to the load side.

## 5.4 EXPERIMENTAL PERFORMANCE OF A WAVELET MODULATED INVERTER

The preceding section has provided results of simulating the performance of a single-phase voltage-source $H$-bridge wavelet modulated (WM) inverter for supplying different load types. These simulation results have demonstrated the WM inverter's high ability to substantially eliminate harmonics from output voltages and currents. This section presents the real-time implementation and experimental performance of a $1\phi$ VS $H$-bridge IGBT WM inverter when supplying different load types. The procedure for implementing the wavelet modulation technique is realized using a *Turbo-C* code. When the developed *Turbo-C* code is executed using the dSPACE ds1104 DSP board, switching pulses are generated and sent to the digital output port (DO/P) of the DSP board. These switching pulses are fed to a pulse amplifier before being applied to the gates of the IGBT switching elements. In addition, hall-effect sensors are used to collect currents and voltages for performing spectral analysis using a Tektronix TDS 3014B 4-channel digital phosphor oscilloscope.

The tested 5 kW IGBT VS inverter output voltages and load currents are collected using the hall-effect sensors, displayed, accessed, and analyzed using a Tektronix TDS 3014B four-channel digital phosphor oscilloscope. Four types of loads are tested to investigate the experimental performance of the WM inverter. The tested loads are as follows:

- Linear load: a resistive load with $R_L = 15\,\Omega$
- Linear load: an $R - L$ load with an impedance $Z_L = 15 + j12.88\,\Omega$
- Dynamic load: a $1\phi$, $\frac{1}{2}\,hp$, 60 Hz, 120 V, 1750 RPM capacitor-run induction motor
- Nonlinear Load: A $1\phi$ full-wave diode rectifier with a resistive load of $25\,\Omega$

### *The Resistive Load (Linear Load)*

The first experimental performance test is conducted for the IGBT inverter to supply a linear load with a voltage at frequency of 60 Hz. The IGBT VS inverter has a dc input voltage of 50 V. The inverter output voltage (load voltage) $V_o(t)$ and load current $I_L(t)$ are collected using a hall-effect sensors in order to determine their spectra and total harmonic distortion factors. The collected load voltage and current are fed into the Tektronix TDS 3014B four-channel digital phosphor oscilloscope to determine the spectra of $V_o(t)$ and $I_L(t)$. Figure 5.11 shows the WM inverter output voltage $V_o(t)$ and the resistive load current $I_L(t)$. The spectra of the WM inverter output voltage $V_o(t)$ and the load current are determined and shown in Figure 5.12.

The voltage and current spectra shown in Figure 5.12 confirm the improvements in the quality of the inverter outputs observed in simulation results. Also, these spectra show that significant amounts of energy are concentrated in the voltage and current fundamental components, while almost negligible energy is allocated in

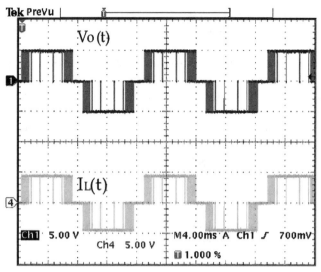

**Figure 5.11**   The resistive load: the WM inverter output voltage $V_o(t)$ and the load current $I_L(t)$. $V_o(t)$ scale: 50 V/Div., $I_L(t)$ scale: 4 A/Div., and time scale: 4 m. sec./Div.

harmonic components, which facilitates transferring high power from the supply to the load.

### The R – L Load (Linear Load)

The experimental performance for the $R - L$ load is carried out by activating the 5 kW $1\phi$ voltage source $H$-bridge IGBT inverter using switching pulses generated by the wavelet modulation technique. The switching pulses are generated and sent to the DAC of the DSP board, where they are applied to the gates of IGBT switches through a pulse amplifier. The dc input voltage is set to $V_{DC} = 50\,V$, and the $R - L$ load of $10 + j7.36\,\Omega$ is connected to the inverter output side. Note that this test is carried out in order to investigate the open-loop steady-state performance of the proposed WM inverter when supplying a static $R - L$ load. Figure 5.13 shows experimental voltage $V_o(t)$ and load current $I_L(t)$ for the WM inverter. The magnitude spectra of the experimental voltage $V_o(t)$ and load current $I_L(t)$ of the WM inverter are determined for this test. The voltage and current magnitude spectra are shown in Figure 5.14.

The spectra of voltage and current in Figure 5.14 confirm the improvements in the quality of the inverter outputs observed in the simulation results. Also, these spectra show that significant amounts of energy are concentrated in the voltage and current fundamental components, while almost negligible energy is allocated in harmonic components, which facilitates transferring high power from the supply to the load.

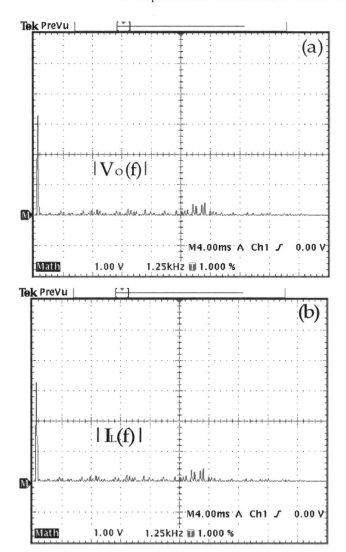

**Figure 5.12** The resistive load: (a) the magnitude spectrum of the WM inverter output voltage $V_o(t)$ and (b) the magnitude spectrum of the load current $I_L(t)$. $|V_o(f)|$ scale: 10 V/Div., $|I_L(f)|$ scale: 1 A/Div., and frequency scale: 1.25 kHz/Div.

### The Single-Phase Capacitor-Run Induction Motor (Dynamic Load)

A single-phase induction motor is investigated for the experimental performance of the WM inverter when supplying the dynamic load. The motor terminals are connected to the 5 kW *H*-bridge IGBT inverter output terminals. The inverter is activated using switching pulses generated by the wavelet modulation technique with an input dc voltage of 170 V. The inverter output voltage $V_o(t)$ and the motor

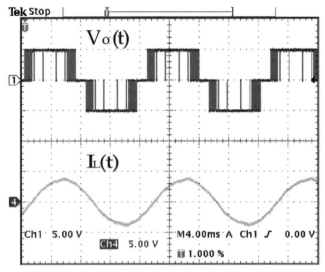

**Figure 5.13**    The $R - L$ load: the WM inverter output voltage $V_o(t)$ and the load current $I_L(t)$. $V_o(t)$ scale: 50 V/Div., $I_L(t)$ scale: 4 A/Div., and time scale: 4 m. sec./Div.

current $I_m(t)$ of the WM inverter are shown in Figure 5.15. Figure 5.16 shows the magnitude spectra of the experimental output voltage $V_o(t)$ and motor current $I_m(t)$ of the WM inverter. Figure 5.16 (a) shows the magnitude of the fundamental output voltage $V_1 = 82\,V$ RMS. Note that the voltage harmonics of some significant magnitudes are located in high frequency bands. Figure 5.16 (b) shows that the current harmonics are quite reduced.

### The Diode Rectifier (Nonlinear Load)

The nonlinear load performance test is carried out through supplying a $1\phi$ full-wave diode rectifier that has a resistive load of $R = 25\,\Omega$. The WM inverter has a dc input voltage of 50 V. The inverter output voltage $V_o(t)$ and the load current $I_L(t)$ are collected and analyzed. Figure 5.17 shows the WM inverter output voltage $V_o(t)$ and the nonlinear load current $I_L(t)$. Figure 5.18 shows the magnitude spectra of the experimental output voltage $V_o(t)$ and the nonlinear load current $I_L(t)$ of the WM inverter.

The spectra of the $1\phi$ full-wave diode rectifier input voltage and current shown in Figures 5.18 demonstrate again the improvements in the quality of the inverter output when supplying a nonlinear load. Moreover, these spectra show one more time the significant amounts of energy concentrated in the voltage and current fundamental components, while almost negligible energy is allocated in harmonic components.

Simulation and experimental performance results presented in the preceding two sections have shown consistent performance of the WM inverter when supplying

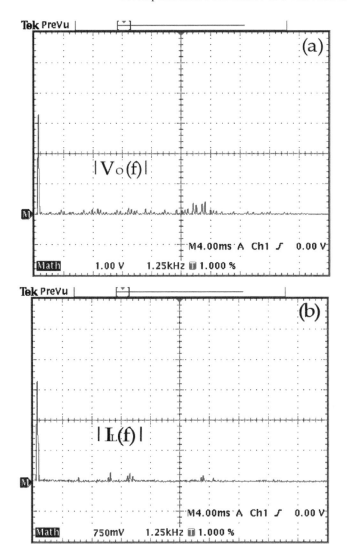

**Figure 5.14**    The $R - L$ load: (a) the magnitude spectrum of the WM inverter output voltage $V_o(t)$ and (b) the magnitude spectrum of the load current $I_L(t)$. $|V_o(f)|$ scale: 10 V/Div., $|I_L(f)|$ scale: 0.75 A/ Div., and frequency scale: 1.25 kHz/Div.

linear, dynamic and nonlinear loads. Table 5.1 summarizes the results of all simu-lated and experimental tests.

The data in Table 5.1 show a high degree of consistency between simulated and experimental tests. The conclusion that can be drawn from these consistent results is that the wavelet modulated technique is not sensitive to the implementation using different platforms. Also, the wavelet modulation technique can accurately provide switching elements with their needed ON and OFF switching times without a need

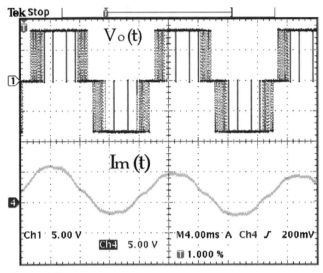

**Figure 5.15**   The $1\phi$ induction motor: the WM inverter output voltage $V_o(t)$ and the motor current $I_m(t)$. $V_o(t)$ scale: 100 V/Div., $I_L(t)$ scale: 1 A/Div., and time scale: 4 m. sec./Div.

for compensating the locations and/or the durations of each generated switching pulse. In addition, the data in Table 5.1 show the high amount of power that can be transferred from the input side to the output side of the WM inverter. This ability for high power transfer can be linked to the effective switching that improves both magnitudes of output fundamental components and the inverter efficiency.

In order to provide figures of performance comparison between the wavelet modulation technique for $1\phi$ $H$-bridge voltage source inverters and both the pulse-width modulation (PWM) and the random pulse-width modulation (RPWM) techniques, Table 5.2 summarizes the magnitudes of output voltage fundamental components ($|V_1|$, voltage total harmonic distortion (THD$_V$) factors, memory requirements, and execution time ($T_{exec.}$) for the three modulation techniques. The magnitudes of the output voltage fundamental components have been normalized with respect to the inverter dc input voltage. Furthermore, the memory sizes and execution times are determined based on experimental implementation and real-time execution of the three modulation techniques using the dSPACE ds 1104 DSP board.

The data presented in Table 5.2 show that the wavelet modulation technique is capable of offering significant improvements on the output voltage of the $1\phi$ $H$-bridge voltage source inverters over other conventional modulation technique. Also, the data in Table 5.2 show that the wavelet modulation technique requires smaller memory size and shorter time for execution than those required by other modulation techniques due to the fact that wavelet modulation technique does not require generating carrier signals or performing signal comparison steps.

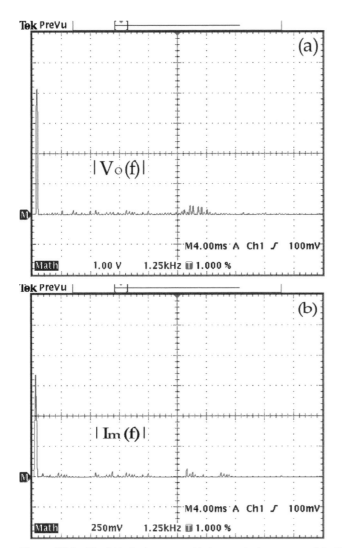

**Figure 5.16**  The $1\phi$ induction motor: (a) the magnitude spectrum of the WM inverter output voltage $V_o(t)$ and (b) the magnitude spectrum of the motor current $I_m(t)$. $|V_o(f)|$ scale: 20 V/Div., $|I_L(f)|$ scale: 0.25 A/Div., and frequency scale: 1.25 kHz/Div.

## 5.5  THE SCALE-TIME INTERVAL FACTOR $\gamma$

Switching pulses generated to operate the inverter represent a set of synthesis scaling functions ($\{\tilde{\varphi}_j(t)\}$) created to reconstruct the reference-modulating signal $S_M(t)$ from its nonuniform recurrent samples. As the scale $j$ changes, the duration and the location of each dilated and translated version of the synthesis scaling function ($\tilde{\varphi}_{j,k}(t)$) change over each half-cycle of $S_M(t)$. However, the scale is constant over

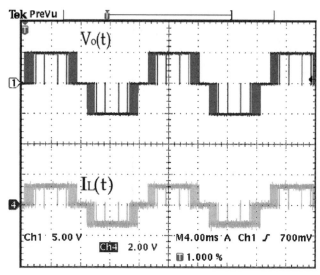

**Figure 5.17**    The nonlinear load: the WM inverter output voltage $V_o(t)$ and the load current $I_L(t)$. $V_o(t)$ scale: 50 V/Div., $I_L(t)$ scale: 4 A/Div., and time scale: 4 m. sec./Div.

**Table 5.1**    Simulation and experimental performance test results of the $1\phi$ VS $H$-bridge WM inverter.

| Load | Parameter | Simulation | Experimental |
|------|-----------|------------|--------------|
| $R$ | $|V_1|$ | 48.24 V | 46.67 V |
| | $|I_1|$ | 3.22 A | 3.10 A |
| | $THD_V$ | 17.68% | 19.26% |
| | $THD_I$ | 17.68% | 19.26% |
| $L-R$ | $|V_1|$ | 48.24 V | 47.87 V |
| | $|I_1|$ | 2.37 A | 2.05 A |
| | $THD_V$ | 17.64% | 19.10% |
| | $THD_I$ | 2.13% | 4.86% |
| Motor | $|V_1|$ | 108.38 V | 113.23 V |
| | $|I_1|$ | 3.1 A | 2.75 A |
| | $THD_V$ | 17.86% | 16.84% |
| | $THD_I$ | 1.69% | 6.16% |
| Rectifier | $|V_1|$ | 46.35 V | 44.38 V |
| | $|I_1|$ | 6.23 A | 5.41 A |
| | $THD_V$ | 14.68% | 16.84% |
| | $THD_I$ | 26.43% | 24.53% |

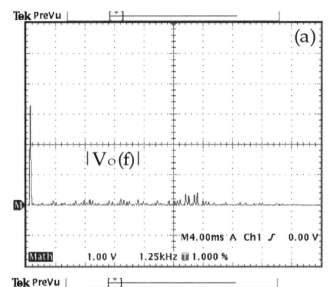

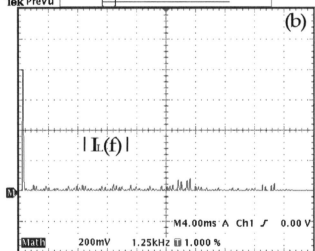

**Figure 5.18**    The nonlinear load: (a) the magnitude spectrum of the WM inverter output voltage $V_o(t)$ and (b) the magnitude spectrum of the nonlinear load current $I_L(t)$. $|V_o(f)|$ scale: 20 V/Div., $|I_L(f)|$ scale: 0.25 A/Div., and frequency scale: 1.25 kHz/Div.

**Table 5.2**    Performance comparison between the WM, PWM, and RPWM inverters.

| Parameter | WM | PWM | RPWM |
|---|---|---|---|
| $|V_1|_{pu}$ | 95.74% | 83.28% | 71.42% |
| $THD_V$ | 19.10% | 51.32% | 40.74% |
| Memory size | 43 kB | 49 kB | 51.6 kB |
| $T_{exec.}$ | 84 $\mu$sec. | 124 $\mu$sec. | 153 $\mu$sec. |

the interval of support of $\tilde{\varphi}_{j,k}(t)$. On the other hand, the changes of the scale $j$ with time show the location of sample groups along with their density. This indication can help to improve the quality of the inverter outputs. Also, the value of the scale $j$ that is associated with the switching pulses of the maximum width can be located. Furthermore, as the scale is increasing, the derivative of the sampled reference-modulating signal is positive, which implies that the sampled signal is changing toward its maximum value.

A factor $\gamma$ can be defined as a *scale–time interval* factor to provide a time-scale relation that can produce time localized values of the scale over a half-cycle of the reference-modulating signal $S_M(t)$. This factor can be defined as follows:

*Definition 5.1:* let $\gamma$ be a *scale–time interval* factor for sample group $d$ defined as:

$$\gamma_d = \gamma_{d-1} + \frac{S_M'(t_{d2})}{|S_M'(t_{d2})|} \sum_{m=1}^{j} m 2^m (t_{m2} - t_{m1}) \qquad (5.10)$$

where $\gamma_d$ is the scale-time interval factor for the sample group $d$ and $j$ is the scale and the value of $t_{d2}$ is given as:

$$t_{d2} = d + 1 - 2^{-(j+1)} \qquad (5.11)$$

For $S_M'(t_{d2}) = 0$, the value of $t_{d2}$ has to be:

$$t_{d2} = \frac{lT_m D}{4} \qquad (5.12)$$

where $l = 1, 3$.
Substituting the value of $t_{d2}$ in equation (5.15) yields:

$$d = \frac{lT_m D}{4} - 1 + 2^{-(j+1)} \qquad (5.13)$$

Calculating $d$ using equation (5.17) for $D = 30$, $l = 1, 3$, and $j = 1, 2,...$ produces fractional numbers, which contradict the fact that $d \in \mathbb{Z}$. As a result:

$$t_{d2} \neq \frac{lT_m D}{4} \Rightarrow S_M'(t_{d2}) \neq 0 \qquad (5.14)$$

The factor $\gamma$ changes its increasing pattern into a decreasing one following a change in the sign of the first derivative of the reference modulating signal $S_M'(t)$. As a consequence, the term $\dfrac{S_M'(t_{d2})}{|S_M'(t_{d2})|}$ ensures changing $k$ from an increasing sequence to a decreasing one follows any change in the sign of $S_M'(t)$. This can be very important when relating the change in the scale to the change in a sampled signal. Figure 5.19 shows the scale–time interval factor $\gamma$ for the WM inverter output voltage at a frequency of 60 Hz. The scale–time interval factor $\gamma$ can provide instantaneous values of time durations for each interval of support for any synthesis function over one cycle of $S_M(t)$. Such values can be used to change locations of

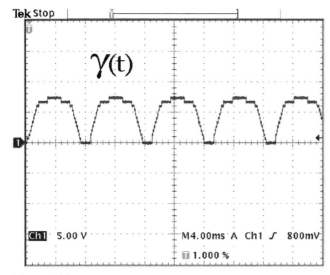

**Figure 5.19**    The scale–time interval factor $\gamma$ for the WM inverter output voltage at a frequency of $f = 60\,\text{Hz}$.

sample groups that can be employed to adjust magnitudes as well as frequencies of inverter outputs. As a result, changing intervals of support for dilated and shifted versions of the synthesis scaling function ($\tilde{\varphi}_{j,k}(t)$) can change durations as well as locations of ON switching pulses.

## *Example*

A single-phase ($1\phi$) voltage source (VS) $H$-bridge WM inverter supplies an $R - L$ load with an impedance $\bar{Z}_L = 5 + j12\,\Omega$. If the frequency of the output voltage is $f_o = 60\,\text{Hz}$, the input dc voltage $V_{DC} = 90\,\text{V}$, and the maximum value of the scale is $J = 12$, calculate:

1. The rms value of the fundamental frequency component of the output voltage $|V_1|_{rms}$
2. The rms value of the fundamental frequency component of the load current $|I_1|_{rms}$
3. The switch utilization ratio $F_{UR}$
4. The maximum value of the scale-time interval factor $\gamma_{max}$

## *Solution*

1. $|V_1|_{rms}$ is calculated using the data of Table 5.2 as

$$|V_1|_{rms} = \frac{0.9547 \times V_{DC}}{\sqrt{2}} = \frac{0.9547 \times 90}{\sqrt{2}} = 60.76\,\text{V}$$

2. The rms value of the fundamental frequency component of the load current $|I_1|_{rms}$ is determined as:

$$|I_1|_{rms} = \frac{|V_1|}{|Z_L|} = \frac{60.76}{\sqrt{5^2 + 13^2}} = 4.36 \, A$$

3. The switch utilization ratio $F_{UR}$ is calculated as:

$$F_{UR} = \frac{|V_1| \times |I_1|}{q \times V_{DC} \times |I_1|_{peak}} \tag{5.15}$$

where $q$ is the number of the inverter switches, which for the case of a H-bridge inverter $q = 4$. The switch utilization ratio $F_{UR}$ is:

$$F_{UR} = \frac{60.76 \times 4.36}{4 \times 90 \times \sqrt{2} \times 4.36} = 0.12$$

4. The value of the scale–time interval factor $\gamma$ can be calculated using equation (5.10). The maximum value of the factor $\gamma$ occurs when the scale $j$ is maximum (i.e., for $j = J$). This can be stated as:

$$(\gamma_d)_{max} = \sum_{m=1}^{J-1} \gamma_m + \frac{S_M'(t_{J2})}{|S_M'(t_{J2})|} J2^J (t_{J2} - t_{J1}) \tag{5.16}$$

Evaluating the factor $\gamma_d$, $t_{d1}$, $t_{d2}$ and $S_M'(t_{d2})$ for $d = 1, 2..., 30$ and $j = 1, 2,...,$ $J$ produces $\gamma_{max} = 9.06$. Note that $t_{d1}$ and $t_{d2}$ are calculated using the following expressions as:

$$t_{d1} = d + 2^{-(j+1)}$$

$$t_{d2} = d + 1 - 2^{-(j+1)}, d = 1, 2, ..., \mathcal{D}_\varphi, j = 1, 2 ... \tag{5.17}$$

## 5.6  SUMMARY

The simulation and experimental performances of the $1\phi$ wavelet modulated (WM) inverter show significantly high-quality output voltages and load currents. This high quality is demonstrated through the high magnitudes of fundamental components as well as low harmonic contents of the output voltages and currents of the WM inverter when supplying linear, dynamic, and nonlinear loads. Moreover, performance test results show that the WM inverter is capable of transferring high power to the load side. Furthermore, simulation and experimental test results are consistent, which indicates that the wavelet modulation technique has high degree of insensitivity to implementation platforms. Finally, the scale–time interval factor $\gamma$ is introduced in order to define a localized-time change in the scale $j$ of the synthesis scaling function $\tilde{\varphi}_j(t)$.

# Chapter **6**

# Three-Phase Wavelet Modulated Inverters

## 6.1  INTRODUCTION

The last chapter presented performance simulation and experimental test results of single-phase ($1\phi$) WM inverters for supplying different types of loads. These test results demonstrate significant capabilities to produce high-quality outputs and transfer more power to the load side, along with a high degree of insensitivity to the implementation platform. In general, inverters can be designed as single-phase or three-phase ($3\phi$) configurations. Three-phase inverters have gained popularity in various industrial applications, including motor drives, power systems, and renewable and non-renewable energy utilization systems. Continuous research on and development of the function and integration of inverters in industrial applications represent evidence for the increasing demands for these devices. The technical literature reports several modulation techniques that have been developed and successfully tested for operating both $1\phi$ and $3\phi$ inverters. These modulation techniques have been developed to optimize the functions of inverters by achieving:

1. Wide range of linear modulation
2. Minimum switching losses
3. High magnitude of the fundamental output frequency component
4. Low output harmonic contents
5. Simple implementation for practical applications

This chapter presents the implementation of nondyadic-type multiresolution analyses (MRAs) that can be employed to generate switching pulses for operating a $3\phi$ six-pulse inverter. The desired nondyadic MRA is constructed using the scale-based linearly combined basis functions, which are presented in Chapter 4.

*An Introduction to Wavelet Modulated Inverters*, by S.A. Saleh and M. Azizur Rahman
Copyright © 2011 Institute of Electrical and Electronics Engineers

## 6.2 IMPLEMENTING THE WAVELET MODULATION TECHNIQUE FOR A THREE-PHASE INVERTER

The basis of the wavelet modulation technique lies in the definition of the three shifted scale-based linearly combined scaling functions, which are used to sample the reference modulating signals in a nonuniform recurrent manner. In a $3\phi$ six-pulse inverter, three reference-modulating signals are used to generate the required switching pulses. These three reference-modulating signals have the same frequency and magnitude with a phase shift of $\dfrac{2\pi}{3}$ from each other as:

$$S_{Ma}(t) = \sin(\omega_m t) \tag{6.1}$$

$$S_{Mb}(t) = \sin\left(\omega_m t - \frac{2\pi}{3}\right) \tag{6.2}$$

$$S_{Mc}(t) = \sin\left(\omega_m t + \frac{2\pi}{3}\right) \tag{6.3}$$

In Chapter 4, the three scale-based linearly combined scaling functions have been defined at scale $j$ as:

$$\left(\varphi_a(t)\right)_j = \phi\left(2^{j+1}t\right) + \phi\left(2^{j+1}(t-u)\right) \tag{6.4}$$

$$\left(\varphi_b(t)\right)_j = \phi\left(2^{j+1}t - z_b\right) + \phi\left(2^{j+1}(t-u) - z_b\right) \tag{6.5}$$

$$\left(\varphi_c(t)\right)_j = \phi\left(2^{j+1}t - z_c\right) + \phi\left(2^{j+1}(t-u) - z_c\right) \tag{6.6}$$

where $u = 1 - 2^{-(j+1)}$, $j = 1, 2,\dots$ and the shifts $z_b$ and $z_c$ are given by:

$$z_b = \frac{2^{2-j}}{3} \tag{6.7}$$

$$z_c = \frac{2^{3-j}}{3} \tag{6.8}$$

The shifts $z_b$ and $z_c$ are dependent on the scale $j$ due to the structure of $\varphi(t)$ as well as the fact that one group of samples is created at each scale. Appendix A provides detailed mathematical derivations of $z_b$ and $z_c$. The three scaling functions $\varphi_a(t)$, $\varphi_b(t)$, and $\varphi_c(t)$ are capable of creating sets of nonuniform recurrent sample groups of the three reference-modulation signals $S_{Ma}(t)$, $S_{Mb}(t)$, and $S_{Mc}(t)$. The sampling of three continuous-time (CT) reference modulating signals using different sampling functions requires three synthesis functions to reconstruct these CT signals from their samples. Reconstructing the three CT reference modulation signals $S_{Ma}(t)$, $S_{Mb}(t)$, and $S_{Mc}(t)$ from samples created by $(\varphi_a(t))_{(j,k)}$, $(\varphi_b(t))_{(j,k)}$, and $\varphi_c(t))_{(j,k)}$ is carried out by dilated and translated version of their dual synthesis scaling functions. These scale-based linearly combined synthesis scaling functions are defined in Chapter 4 as:

$$\left(\tilde{\varphi}_a(t)\right)_j = \phi_H(t) - \left(\varphi_a(t)\right)_j \tag{6.9}$$

$$\left(\tilde{\varphi}_b(t)\right)_j = \phi_H(t - z_b) - \left(\varphi_b(t)\right)_j \tag{6.10}$$

$$\left(\tilde{\varphi}_c(t)\right)_j = \phi_H(t - z_c) - \left(\varphi_c(t)\right)_j \tag{6.11}$$

where $j = 1, 2, \ldots$.

The nondyadic type MRA representation of the three CT reference modulating signals provides a mathematical expression for the output voltage of each leg of a $3\phi$ voltage source WM inverter as:

$$\frac{v_a(t)}{V_{DC}} = \sum_{j=1}^{J} \left(\tilde{\varphi}_a(t)\right)_j - \left(\tilde{\varphi}_a\left(t - \frac{T_m}{2}\right)\right)_j + \sum_{j=1}^{J-1} \left(\tilde{\varphi}_a(t)\right)_{(J-j)} - \left(\tilde{\varphi}_a\left(t - \frac{T_m}{2}\right)\right)_{(J-j)} \tag{6.12}$$

$$\frac{v_b(t)}{V_{DC}} = \sum_{j=1}^{J} \left(\tilde{\varphi}_b(t)\right)_j - \left(\tilde{\varphi}_b\left(t - \frac{T_m}{2}\right)\right)_j + \sum_{j=1}^{J-1} \left(\tilde{\varphi}_b(t)\right)_{(J-j)} - \left(\tilde{\varphi}_b\left(t - \frac{T_m}{2}\right)\right)_{(J-j)} \tag{6.13}$$

$$\frac{v_c(t)}{V_{DC}} = \sum_{j=1}^{J} \left(\tilde{\varphi}_c(t)\right)_j - \left(\tilde{\varphi}_c\left(t - \frac{T_m}{2}\right)\right)_j + \sum_{j=1}^{J-1} \left(\tilde{\varphi}_c(t)\right)_{(J-j)} - \left(\tilde{\varphi}_c\left(t - \frac{T_m}{2}\right)\right)_{(J-j)} \tag{6.14}$$

where $v_a(t)$, $v_b(t)$, and $v_c(t)$ are the voltages across the three inverter legs, $V_{DC}$ is the inverter input dc voltage, and $J = \max(j)$.

The three CT reference-modulating signals are processed using the nondyadic type MRAs constructed using $\varphi_a(t)$, $\varphi_b(t)$, $\varphi_c(t)$, $\tilde{\varphi}_a(t)$, $\tilde{\varphi}_b(t)$, and $\tilde{\varphi}_c(t)$. The three synthesis scaling functions $\tilde{\varphi}_a(t)$, $\tilde{\varphi}_b(t)$, and $\tilde{\varphi}_c(t)$ provide switching signals for the $3\phi$ inverter. The implementation of the wavelet modulation technique for a $3\phi$ voltage source six-pulse inverter can be divided into two parts as follows:

- Sampling $S_{Ma}(t)$, $S_{Mb}(t)$, and $S_{Mc}(t)$ in a nonuniform recurrent manner using the $\{(\varphi_a(t))_{(j,k)}\}$, $\{(\varphi_b(t))_{(j,k)}\}$, and $\{(\varphi_c(t))_{(j,k)}\}$ sets of wavelet basis functions. The created samples in sample groups $da$, $db$, and $dc$ at scale $j$ for the three CT sampled signals are located at the boundaries of the intervals $[t_{da1}, t_{da2}]$, $[t_{db1}, t_{db2}]$, and $[t_{dc1}, t_{dc2}]$. These time locations are determined as:

$$\underbrace{\begin{aligned} t_{da1} &= d + 2^{-j-1} \\ t_{da2} &= d + 1 - 2^{-j-1} \end{aligned}}_{\text{for } S_{Ma}(t)} \tag{6.15}$$

$$\underbrace{\begin{aligned} t_{db1} &= z_b(j) + d + 2^{-j-1} \\ t_{db2} &= z_b(j) + d + 1 - 2^{-j-1} \end{aligned}}_{\text{for } S_{Mb}(t)} \tag{6.16}$$

$$\underbrace{\begin{aligned} t_{dc1} &= z_c(j) + d + 2^{-j-1} \\ t_{dc2} &= z_c(j) + d + 1 - 2^{-j-1} \end{aligned}}_{\text{for } S_{Mc}(t)} \tag{6.17}$$

where $d = 0, 1, 2, \ldots, \mathcal{D} - 1$, $\mathcal{D}$ is the number of created sample groups over $T_m$ and $j = 1, 2, \ldots$.

- Generating switching pulses using the three synthesis scale-based linearly combined scaling function $(\tilde{\varphi}_a(t))_{(j,k)}$, $(\tilde{\varphi}_b(t))_{(j,k)}$, and $(\tilde{\varphi}_c(t))_{(j,k)}$, where:

$$(\tilde{\varphi}_a(t))_{(j,k)} = \tilde{\varphi}_a(2^j t - k), \quad j = 1, 2, \ldots, k \in \mathbb{Z}$$

$$(\tilde{\varphi}_b(t))_{(j,k)} = \tilde{\varphi}_b(2^j t - k), \quad j = 1, 2, \ldots, k \in \mathbb{Z}$$

$$(\tilde{\varphi}_c(t))_{(j,k)} = \tilde{\varphi}_c(2^j t - k), \quad j = 1, 2, \ldots, k \in \mathbb{Z}$$

Each switching pulse is created as a dilated version at different value of the scale $j$ and translated at different value of the shift $k$ of the three synthesis scaling functions.

The time interval and location of each sample group of each CT reference modulating signal change due to the changes of the scale $j$ and the translation $k$. Note that the starting value of the scale $j$ for each scaling function has to be initialized as $j_0 > 0$. This condition is necessary to avoid uniform sampling with $j_0 = 0$. The wavelet modulation technique for a $3\phi$ inverter can be implemented through the following steps:

- **Step 1:** Set the scales $j_a$, $j_b$, and $j_c$ to $j_0 = 1$, $t = 0$ and the index of sample groups $d$ to $d_0 = 0$ for each phase.
- **Step 2:** Create samples at $t_{da1}$, $t_{da2}$, $t_{db1}$, $t_{db2}$, $t_{dc1}$, and $t_{dc2}$.
- **Step 3:** Verify:
    - **(i).** If $S'_{Ma}(t_{da2}) > 0$, then $j_a = j_a + 1$.
    - **(ii).** If $S'_{Ma}(t_{da2}) < 0$, then $j_a = j_a - 1$.
    - **(iii).** If $S'_{Mb}(t_{db2}) > 0$, then $j_b = j_b + 1$.
    - **(iv).** If $S'_{Mb}(t_{db2}) < 0$, then $j_b = j_b - 1$.
    - **(v).** If $S'_{Mc}(t_{dc2}) > 0$, then $j_c = j_c + 1$.
    - **(vi).** If $S'_{Mc}(t_{dc2}) < 0$, then $j_c = j_c - 1$.

    where:

$$S'_{Ma}(t) = \frac{dS_{Ma}(t)}{dt}$$

$$S'_{Mb}(t) = \frac{dS_{Mb}(t)}{dt}$$

$$S'_{Mc}(t) = \frac{dS_{Mc}(t)}{dt}$$

- **Step 4:** Generate an ON pulse over the time interval of the samples group $d$ for each CT reference-modulating signal.
- **Step 5:** Verify:
    - **(i).** If $t \geq T_m$ set $j_a = 1$ and $da = 0$, else $da = da + 1$.
    - **(ii).** If $t \geq \frac{4}{3} T_m$ set $j_b = 1$ and $db = 0$, else $db = db + 1$.
    - **(iii).** If $t \geq \frac{5}{3} T_m$ set $j_c = 1$ and $dc = 0$, else $dc = dc + 1$.

- **Step 6:** $t = t + T_{st}$; $T_{st}$ is the time-step.
- **Step 7:** Go to step 2.

These steps of implementing the proposed wavelet modulation technique are given in a flowchart of Figure 6.1 in order to facilitate realizing the wavelet modulation technique for performance simulation of a $3\phi$ voltage source six-pulse inverter.

## 6.3 SIMULATED PERFORMANCE OF A THREE-PHASE WAVELET MODULATED INVERTER

The step-by-step procedure to implement the wavelet modulation technique for a three-phase ($3\phi$) voltage source (VS) six-pulse inverter is realized using a Mat-lab code that generates switching pulses to activate a Simulink model of a $3\phi$ VS six-pulse inverter. Four different load types are used to illustrate the simulated performances of the $3\phi$ VS six-pulse wavelet modulated (WM) inverter. The tested load types are:

- Linear load: a $3\phi$ Y-connected resistive load with a resistance of $R_L = 15\,\Omega$/phase
- Linear load: a $3\phi$ Y-connected $R - L$ load with an impedance of $Z_L = 15 + j12.82\,\Omega$/phase
- Dynamic load: a $3\phi$, 1-$hp$, 60 Hz, 208 V, 1746 rpm, Y-connected squirrel-cage induction motor
- Nonlinear load: a $3\phi$ full-wave diode rectifier with a resistive load of $15\,\Omega$

The quality of inverter output voltage and load current are commonly expressed in terms of the total harmonic distortion (THD) factors that have been defined in equations (5.8) and (5.9) of Section 5.3.

### *The Resistive Load (Linear Load)*

The procedure to implement the wavelet modulation technique for a $3\phi$ inverter is realized using MATLAB/Simulink to simulate the performance of a $3\phi$ VS six-pulse inverter. The simulated inverter is activated by switching pulses generated by the wavelet modulation technique and has an input dc voltage of 50 V to supply the Y-connected resistive load. This simulated test is meant to investigate the open-loop steady-state performance of the $3\phi$ wavelet modulated (WM) inverter when supplying a linear load. Figure 6.2 shows the schematic diagram of the $3\phi$ VS six-pulse IGBT inverter and the resistive load. Figure 6.3 shows the $3\phi$ line-to-line voltages $V_{AB}(t)$, $V_{BC}(t)$, and $V_{CA}(t)$ for the resistive load. The $3\phi$ line-to-neutral voltages $V_{AN}(t)$, $V_{BN}(t)$, and $V_{CN}(t)$ for the resistive load are shown in Figure 6.4. Figure 6.5 shows the $3\phi$ line currents drawn by the $3\phi$ Y-connected resistive load.

The output line voltage $V_{AB}(t)$ and the load line current $I_A(t)$ are analyzed using a Matlab built-in FFT function to determine their harmonic spectra and evaluate their $\text{THD}_V$ and $\text{THD}_I$ factors. Figure 6.6 shows the spectra of $V_{AB}(t)$ and $I_A(t)$.

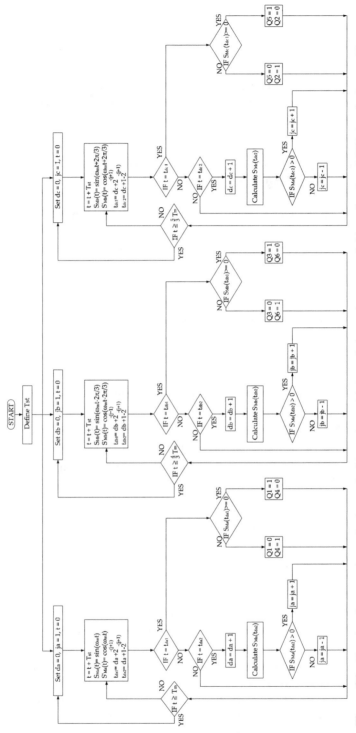

**Figure 6.1** Flowchart for a procedure to implement the $3\phi$ wavelet modulation technique and generate switching pulses for a $3\phi$ VS six-pulse inverter. The time-step $T_{st} = 40\,\mu sec$.

112

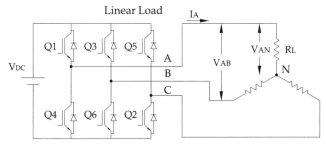

**Figure 6.2**    Schematic diagram of a $3\phi$ VS six-pulse IGBT inverter and the $3\phi$ Y-connected resistive load. $V_{DC} = 50\,\text{V}$.

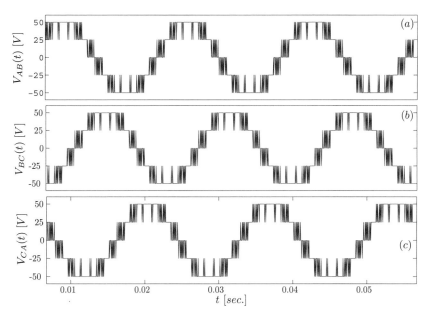

**Figure 6.3**    The $3\phi$ WM inverter with the resistive load: (a) the line voltage $V_{AB}(t)$, (b) the line voltage $V_{BC}(t)$, and (c) the line voltage $V_{CA}(t)$.

The voltage and current spectra in Figure 6.6 show that significant amounts of energy are concentrated in the voltage and current fundamental components, while almost negligible energy is allocated to the harmonic components. These features observed in the voltage and current indicate the ability of the WM inverter to transfer high amounts of power from the supply to the load.

### The R – L Load (Linear Load)

This simulated performance test of the $3\phi$ WM inverter is carried out by connecting an inductive $R - L$ of an impedance of $Z_L = 15 + j12.88\,\Omega$ on the output of the

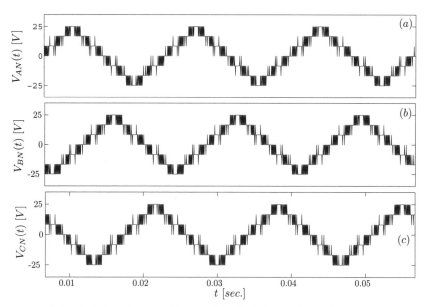

**Figure 6.4**    The 3$\phi$ WM inverter with the resistive load: (a) the phase voltage $V_{AN}(t)$, (b) the phase voltage $V_{BN}(t)$, and (c) the phase voltage $V_{CN}(t)$.

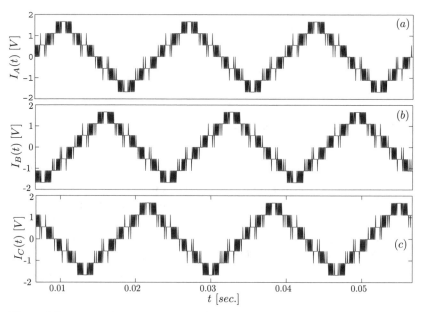

**Figure 6.5**    The 3$\phi$ WM inverter with the resistive load: (a) the line current $I_A(t)$, (b) the line current $I_B(t)$, and (c) the line current $I_C(t)$.

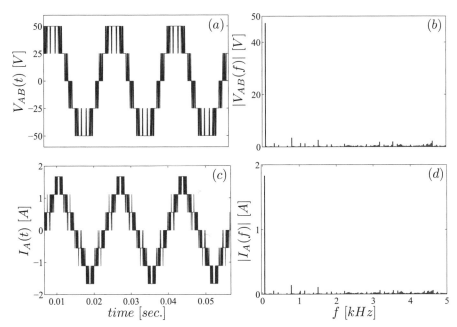

**Figure 6.6**    The $3\phi$ WM inverter with the resistive load: (a) the output line voltage $V_{AB}(t)$, (b) the spectrum of $V_{AB}(t)$, (c) the resistive load current $I_A(t)$, and (d) the the spectrum of $I_A(t)$. $|V_1| = 48.86\,V$, $\text{THD}_V = 14.22\%$, $|I_1| = 1.88\,A$, and $\text{THD}_I = 15.31\%$.

inverter as shown in Figure 6.7. The $R - L$ load simulated test is conducted to investigate the open-loop steady-state performance of the WM inverter when supplying a linear load. Figure 6.8 shows the output line voltage $V_{AB}(t)$ and line current $I_A(t)$ along with their spectra for the $R - L$ load.

The voltage and current spectra in Figures 6.8 show better performances in terms of the magnitude and the harmonic contents of the inverter outputs. The output voltages have low $\text{THD}_V$ factors with the line-to-line fundamental component magnitudes close to the input dc voltage. Moreover, line currents have negligible $\text{THD}_I$ factor values with almost sinusoidal waveforms. These voltage and current harmonic spectra show that the WM inverter has a significant ability to transfer higher power from the supply to the load.

### The $3\phi$ Induction Motor (Dynamic Load)

This simulated test aims to investigate the open-loop steady-state performance of the developed $3\phi$ wavelet modulation technique for activating the $3\phi$ VS six-pulse inverter with an input dc voltage of $300\,V$ to supply a dynamic load. A $1\text{-}hp$, $3\phi$, $208\,V$, 4-pole, $60\,Hz$, $Y$-connected squirrel-cage induction motor is used as the dynamic load as shown in Figures 6.9. Figure 6.10 shows the inverter output voltage $V_{AB}(t)$ and motor line current $I_A(t)$ along with their spectra, when the inverter is

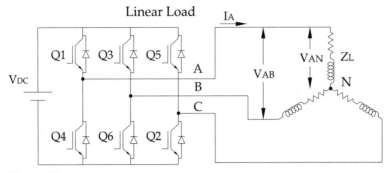

**Figure 6.7**   Schematic diagram of a 3$\phi$ VS six-pulse IGBT inverter and the 3$\phi$ Y-connected $R - L$ load. $V_{DC} = 50\,V$.

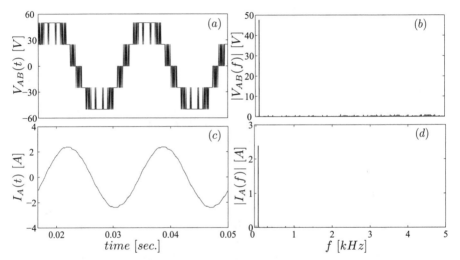

**Figure 6.8**   The 3$\phi$ WM inverter with the $R - L$ load: (a) the WM inverter output voltage $V_{AB}(t)$, (b) the spectrum of $V_{AB}(t)$, (c) the load current $I_A(t)$, and (d) the spectrum of $I_A(t)$. $|V_1| = 48.86\,V$, THD$_V = 14.22\%$, $|I_1| = 2.14\,A$, and THD$_I = 2.13\%$.

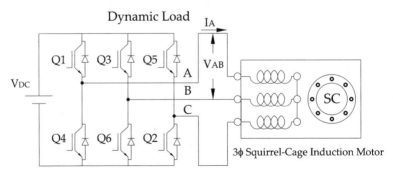

**Figure 6.9**   Schematic diagram of a 3$\phi$ VS six-pulse IGBT inverter and the 1-$hp$, 3$\phi$, 208 V, 4-pole, 60 Hz, Y-connected squirrel-cage induction motor. $V_{DC} = 300\,V$.

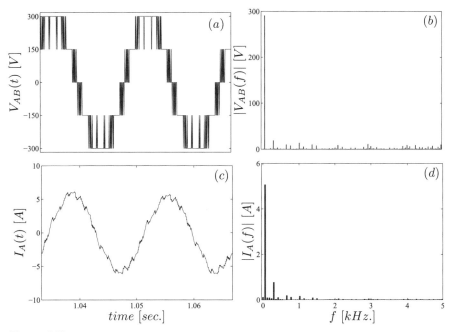

**Figure 6.10**    The $3\phi$ WM inverter with the $3\phi$ induction motor: (a) the WM inverter output line voltage $V_{AB}(t)$, (b) the spectrum of $V_{AB}(t)$, (c) the line current $I_A(t)$, and (d) the spectrum of $I_A(t)$. $|V_1| = 282.94\,V$, THD$_V$ = 27.46%, $|I_1|$ = 5.28 A, and THD$_I$ = 12.78%.

activated using the switching pulses generated by the $3\phi$ wavelet modulation technique.

The inverter output voltage and motor current spectra shown in Figure 6.10 confirm that the $3\phi$ WM inverter is able to provide output voltage with higher magnitudes to supply the $3\phi$ induction motor. Note that the magnitude of the fundamental voltage of the WM inverter is close to the dc input voltage and the harmonic contents are low. The motor current spectra show that the motor current supplied by the WM inverter has low harmonic contents, which can provide the induction motor with less pulsating torque and smoother operation.

### The $3\phi$ Diode Rectifier (Nonlinear Load)

The nonlinear load performance simulated test is carried out to investigate the open-loop steady-state performance of the WM inverter when supplying a nonlinear load. The simulated inverter is activated by switching pulses generated by the three-phase wavelet modulation technique, and has an input dc voltage of 50 V to supply the nonlinear load. The nonlinear load used for this test is a $3\phi$ full-wave diode rectifier that has a resistive load of $R = 25\,\Omega$ as shown in Figure 6.11. Figure 6.12 shows the $3\phi$ WM inverter output voltage $V_{AB}(t)$ and the nonlinear load current $I_A(t)$ along with their harmonic spectra.

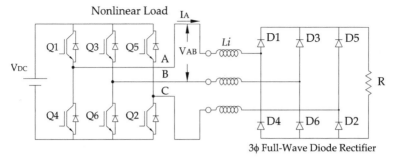

**Figure 6.11**   Schematic diagram of a $3\phi$ VS six-pulse IGBT inverter and the $3\phi$ full-wave diode rectifier. $L_i = 15\,mH$, $R = 25\,\Omega$, and $V_{DC} = 50\,V$.

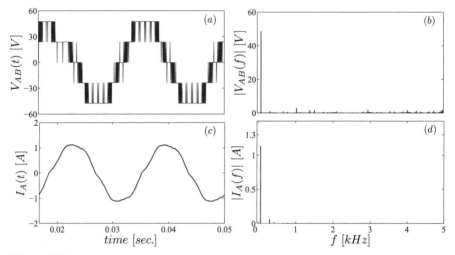

**Figure 6.12**   The $3\phi$ WM inverter with the nonlinear load: (a) the WM inverter output line voltage $V_{AB}(t)$, (b) the spectrum of $V_{AB}(t)$, (c) the line current $I_A(t)$, and (d) the the spectrum of $I_A(t)$. $|V_1| = 47.72\,V$, $THD_V = 15.93\%$, $|I_1| = 1.23\,A$, and $THD_I = 10.31\%$.

The spectra of the $3\phi$ WM inverter output voltage and nonlinear load current shown in Figure 6.12 demonstrate that the output voltage and load current have high fundamental components when supplying a nonlinear load. In addition, the output voltage and load current have low harmonic contents as well as minor fluctuations due to the rectifier ON and OFF switching.

In summary, this section has presented the simulation of open-loop steady-state performances of the proposed wavelet modulation technique for activating a $3\phi$ VS six-pulse inverter to supply different load types. In all performance simulation tests, it has been found that the wavelet modulation technique produced output voltage fundamental components having high magnitudes and low harmonic contents as indicated by evaluating the $THD_V$ and $THD_I$ factors.

## 6.4  EXPERIMENTAL PERFORMANCE OF A THREE-PHASE WAVELET MODULATED INVERTER

The simulation results presented in the preceding section have shown the high ability of the three-phase ($3\phi$) wavelet modulated (WM) inverter to produce high magnitudes of output voltage fundamental components, and to substantially reduce/ eliminate harmonics from output voltages and currents. Also, simulation results have demonstrated that the load type has negligible effects on the output voltage of the $3\phi$ WM inverter. This section presents the real-time implementation and experimental performance of a $3\phi$ voltage source (VS) six-pulse WM inverter when supplying different load types. The procedure for implementing the wavelet modulation technique is realized using a *Turbo-C* code to generate the switching pulses. When the developed *Turbo-C* code is executed using the dSPACE ds1104 DSP board, the generated switching pulses are sent to the digital output port (DO/P) of the DSP board. Furthermore, these switching pulses are fed to a pulse amplifier before being applied to the gates of the inverter switching elements. In addition, hall-effect sensors are used to collect currents and voltages for performing spectral analysis using a Tektronix TDS 3014B 4-channel digital phosphor oscilloscope.

The tested inverter is a 5 kW IGBT VS inverter. The output voltages and load currents are collected using the hall-effect sensors and are displayed, accessed, and analyzed using a Tektronix TDS 3014B four-channel digital phosphor oscilloscope. Four types of loads are tested to investigate the experimental performance of the WM inverter. The tested loads are as the following:

- Linear load: a $3\phi$ Y-connected resistive load with a resistance of $R_L = 15\,\Omega/$ phase
- Linear load: a $3\phi$ Y-connected $R - L$ load with an impedance of $Z_L = 15 + j12.82\,\Omega$/phase
- Dynamic load: a $3\phi$, 1-$hp$, 60 Hz, 208 V, 1746 rpm, Y-connected squirrel-cage induction motor
- Nonlinear load: a $3\phi$ full-wave diode rectifier with a resistive load of $15\,\Omega$

### The Resistive Load (Linear Load)

In this experimental performance test, the IGBT inverter supplies a $3\phi$ Y-connected linear load with a voltage at frequency of 60 Hz. The IGBT VS inverter has a dc input voltage of 50 V. The inverter output line voltage $V_{AB}(t)$ and load line current $I_A(t)$ are collected using hall-effect sensors in order to determine their spectra and total harmonic distortion factors. The collected load voltage and current are fed into the Tektronix TDS 3014B four-channel digital phosphor oscilloscope to determine the spectra of $V_o(t)$ and $I_L(t)$. Figure 6.13 shows the WM inverter output voltage $V_o(t)$ and the resistive load current $I_L(t)$. The spectra of the WM inverter output line voltage $V_{AB}(t)$ and the load line current $I_A(t)$ are determined and shown in Figure 6.14.

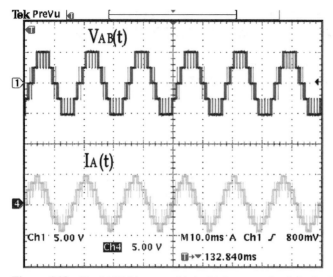

**Figure 6.13** The $3\phi$ WM inverter with the $3\phi$ Y-connected resistive load: the WM inverter output line voltage $V_{AB}(t)$ and the load line current $I_A(t)$. $V_{AB}(t)$ scale: 50 V/Div., $I_L(t)$ scale: 2 A/Div., and time scale: 10 m. sec./Div.

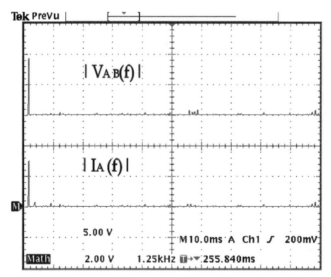

**Figure 6.14** The $3\phi$ WM inverter with the $3\phi$ Y-connected resistive load: the magnitude spectrum of the WM inverter output line voltage $V_{AB}(t)$ and the magnitude spectrum of the load line current $I_A(t)$. $|V_{AB}(f)|$ scale: 20 V/Div., $|I_A(f)|$ scale: 1 A/Div., and frequency scale: 1.25 kHz/Div.

Experimental test results of the $3\phi$ wavelet modulated inverter have shown significant capabilities to eliminate harmonic components from the output voltages. Also, these results have demonstrated clear improvements on the $3\phi$ VS output voltage quality that can be observed from the voltage and current spectra shown in Figure 6.14. Moreover, the fundamental components of the inverter output voltages have peak values close to the input dc voltage indicating effective concentration of output energies in the desired fundamental frequency components. Experimental test results are consistent with simulation ones in both waveform shapes and features of the inverter outputs.

### The R – L Load (Linear Load)

The experimental performance for the $3\phi$ Y-connected $R - L$ load is carried out by activating the 5 kW $3\phi$ VS six-pulse IGBT inverter using switching pulses generated by the wavelet modulation technique. The dc input voltage is set to $V_{DC} = 50\,V$. This test is carried out in order to investigate the open-loop steady-state performance of the $3\phi$ WM inverter when supplying an $R - L$ load. Figure 6.15 shows the experimental $3\phi$ WM inverter output line voltage $V_{AB}(t)$ and $R - L$ load line current $I_A(t)$. The magnitude spectra of the experimental $3\phi$ WM inverter output line voltage $V_{AB}(t)$ and $R - L$ load line current $I_A(t)$ are determined for this test. The voltage and current magnitude spectra are shown in Figure 6.16.

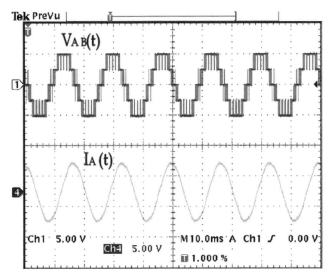

**Figure 6.15**   The $3\phi$ WM inverter with the $3\phi$ Y-connected $R - L$ load: the WM inverter output line voltage $V_{AB}(t)$ and the load line current $I_A(t)$. $V_{AB}(t)$ scale: 50 V/Div., $I_A(t)$ scale: 4 A/Div., and time scale: 10 m. sec./Div.

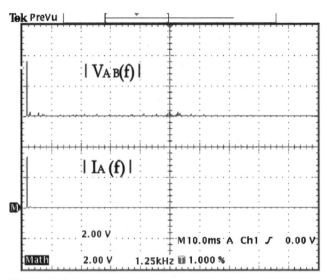

**Figure 6.16**    The $3\phi$ $Y$-connected $R - L$ load: the magnitude spectrum of the WM inverter output voltage $V_o(t)$ and the magnitude spectrum of the load current $I_A(t)$. $|V_{AB}(f)|$ scale: 20 V/Div., $|I_A(f)|$ scale: 1 A/Div., and frequency scale: 1.25 kHz/-Div.

The spectra of output line voltage and load line current in Figure 6.16 confirm the improvements in the quality of the $3\phi$ inverter outputs observed in the simulation results. Also, these spectra show that significant amounts of energy are concentrated in the voltage and current fundamental components, while almost negligible energy is allocated to harmonic components, which facilitates transferring high power from the supply to the load.

### The $3\phi$ Induction Motor (Dynamic Load)

The open-loop steady-state performance of the $3\phi$ VS six-pulse WM inverter is investigated for supplying a dynamic load in this experimental test with an input dc voltage of 300 V. A $3\phi$, 1-$hp$, 60 Hz, 208 V, 1746 rpm, $Y$-connected squirrel-cage induction motor is used as a dynamic load. The motor terminals are connected to the 5 kW six-pulse IGBT inverter output terminals. The inverter is activated using switching pulses generated by the wavelet modulation technique with an input dc voltage of 300 V. The WM inverter output line voltage $V_{AB}(t)$ and the motor line current $I_A(t)$ are shown in Figure 6.17. Figure 6.18 shows the magnitude spectra of the experimental WM inverter output line voltage $V_{AB}(t)$ and motor line current $I_A(t)$.

The experimental test results in Figure 6.18 show the magnitude of the fundamental output line voltage $V_1 = 194\,V$ RMS. Note that the voltage harmonics of some significant magnitudes are located in high frequency bands. Also, Figure 6.18 shows that the current harmonics are quite reduced.

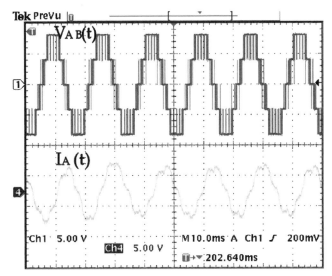

**Figure 6.17**    The $3\phi$ WM inverter with the $3\phi$ induction motor: the WM inverter output line voltage $V_{AB}(t)$ and the motor line current $I_A(t)$. $V_{AB}(t)$ scale: 180 V/-Div., $I_A(t)$ scale: 5 A/Div. and time scale: 10 m. sec./Div.

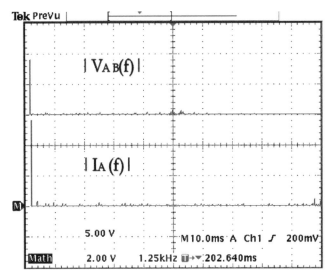

**Figure 6.18**    The $3\phi$ WM inverter with the $3\phi$ induction motor: the magnitude spectrum of the WM inverter output line voltage $V_{AB}(t)$ and the magnitude spectrum of the motor line current $I_A(t)$. $|V_{AB}(f)|$ scale: 100 V/Div., $|I_A(f)|$ scale: 1.75 A/Div., and frequency scale: 1.25 kHz/Div.

## The Diode Rectifier (Nonlinear Load)

This experimental performance test with a nonlinear load is carried out through supplying a $3\phi$ full-wave diode rectifier that has a resistive load of $R = 25\,\Omega$ in the output end. The $3\phi$ VS six-pulse IGBT WM inverter has a dc input voltage of 50 V. The inverter output line voltage $V_{AB}(t)$ and the load line current $I_A(t)$ are collected and analyzed. Figure 6.19 shows the $3\phi$ WM inverter output line voltage $V_{AB}(t)$ and nonlinear load line current $I_A(t)$. Figure 6.20 shows the magnitude spectra of the experimental output line voltage $V_o(t)$ and the nonlinear load line current $I_L(t)$ of the $3\phi$ VS six-pulse IGBT WM inverter.

The spectra of the $3\phi$ full-wave diode rectifier input voltage and current shown in Figures 6.20 demonstrate again the improvements in the quality of the $3\phi$ inverter output when supplying a nonlinear load. Moreover, these spectra show the significant amounts of energy concentrated in the fundamental component of the output voltage, while almost negligible energy is allocated in harmonic components.

Simulation and experimental performance results have shown consistent performance of the WM inverter when supplying linear, dynamic, and nonlinear loads. Table 6.1 summarizes the results of all simulated and experimental tests.

The data in Table 6.1 show a high degree of consistency between simulated and experimental tests conducted to investigate the open-loop steady-state performance of the $3\phi$ VS six-pulse WM inverter. The conclusion that can be drawn from these results is that the three-phase wavelet modulated technique is not sensitive to the implementation using different loads/platforms. Also, the wavelet modulation

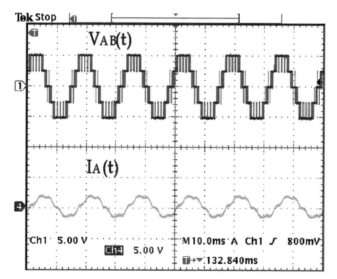

**Figure 6.19**    The $3\phi$ WM inverter with the $3\phi$ nonlinear load: the WM inverter output line voltage $V_{AB}(t)$ and the nonlinear load line current $I_A(t)$. $V_{AB}(t)$ scale: 50 V/Div., $I_A(t)$ scale: 3 A/Div., and time scale: 10 m. sec./Div.

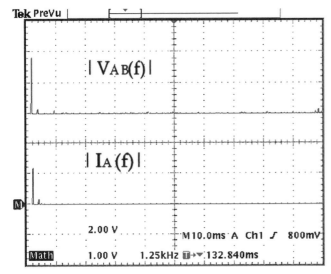

**Figure 6.20**    The 3$\phi$ WM inverter with the 3$\phi$ nonlinear load: the magnitude spectrum of the 3$\phi$ WM inverter output line voltage $V_{AB}(t)$ and the magnitude spectrum of the nonlinear load line current $I_A(t)$. $|V_{AB}(f)|$ scale: 20 V/Div., $|I_A(f)|$ scale: 2 A/Div., and frequency scale: 1.25 kHz/Div.

**Table 6.1**    Simulation and experimental performance test results of the 3$\phi$ VS 6-pulse WM inverter.

| Load | Parameter | Simulation | Experimental |
|---|---|---|---|
| R | $|V_1|$ | 48.86 V | 47.54 V |
| | $|I_1|$ | 1.88 A | 1.63 A |
| | THD$_V$ | 14.22% | 15.29% |
| | THD$_I$ | 15.31% | 16.60% |
| $L - R$ | $|V_1|$ | 48.86 V | 47.15 V |
| | $|I_1|$ | 2.14 A | 2.01 A |
| | THD$_V$ | 14.22% | 15.13% |
| | THD$_I$ | 2.13% | 3.06% |
| Motor | $|V_1|$ | 282.94 V | 274.4 V |
| | $|I_1|$ | 5.28 A | 6.81 A |
| | THD$_V$ | 13.66% | 15.84% |
| | THD$_I$ | 8.78% | 7.16% |
| Rectifier | $|V_1|$ | 47.72 V | 46.18 V |
| | $|I_1|$ | 1.23 A | 1.00 A |
| | THD$_V$ | 15.93% | 19.89% |
| | THD$_I$ | 10.31% | 16.38% |

**Table 6.2** Performance comparison between the WM, PWM, and RPWM inverters.

| Parameter | WM | SVM | RPWM | HBCC |
|---|---|---|---|---|
| $|V_1|_{pu}$ | 96.4% | 87.48% | 72.63% | 77.2% |
| $THD_V$ | 14.22% | 19.45% | 35.87% | 41.64% |
| Memory size | 76 kB | 169 kB | 148 kB | 183 kB |
| $T_{exec}$ | 121 $\mu$sec. | 228 $\mu$sec. | 315 $\mu$sec. | 368 $\mu$sec. |

technique can accurately provide switching elements with their needed ON and OFF switching times without a need for compensating the locations and/or the durations of each generated switching pulse. In addition, the data in Table 6.1 show the high amount of power that can be transferred from the input side to the output side of the $3\phi$ WM inverter. This ability for high power transfer with high magnitude of fundamental components and low harmonic contents can be linked to the effective switching that improves both the magnitude of output fundamental component and the inverter efficiency.

The features of the wavelet modulation technique for $3\phi$ voltage source six-pulse inverters can be better demonstrated by means of performance comparisons with other modulation techniques. Table 6.2 summaries performances of the magnitudes of output voltage fundamental components ($|V_1|$, voltage total harmonic distortion ($THD_V$) factors, memory requirements, and execution time ($T_{exec}$) for the wavelet modulation, space vector modulation (SVM), random pulse-width modulation (RPWM), and hysteresis-band current control (HBCC) techniques. The magnitudes of the output voltage fundamental components have been normalized with respect to the inverter dc input voltage. Furthermore, the memory sizes and execution times are determined based on experimental implementation and real-time execution of the three modulation techniques using the dSPACE ds 1104 DSP board.

The data in Table 6.2 show that the wavelet modulation technique can offer significant improvements on the output voltage of the $3\phi$ voltage source six-pulse inverters over other modulation techniques. In addition, Table 6.2 shows that the wavelet modulation technique requires smaller memory size and shorter time for execution than those required by other modulation techniques due to the fact that wavelet modulation technique does not require generating carrier signals or performing signal comparison steps.

## *Example*

A three-phase ($3\phi$) voltage source (VS) six-pulse WM inverter supplies a $3\phi$, Y-connected $R - L$ load with an impedance $\bar{Z}_L = 10 + j7\,\Omega$/phase. If the frequency of the inverter output voltage is $f_o = 60\,Hz$ and the input dc voltage $V_{DC} = 90\,V$, calculate:

1. The rms value of the fundamental frequency component of the output line voltage $|V_{AB1}|_{rms}$

2. The rms value of the fundamental frequency component of the output phase voltage $|V_{AN1}|_{rms}$

3. The rms value of the fundamental frequency component of the load line current $|I_{A1}|_{rms}$

4. Plot the waveforms of $i_{A1}(t)$ and $v_{AB1}(t)$, and $i_{A1}(t)$ and $v_{AN1}(t)$

### *Solution*

1. $|V_{AB1}|_{rms}$ is calculated using the data from Table 6.2 as:

$$|V_{AB1}|_{rms} = \frac{0.964 \times V_{DC}}{\sqrt{2}} = \frac{0.964 \times 90}{\sqrt{2}} = 61.35 \text{ V}$$

2. $|V_{AN1}|_{rms}$ is determined as:

$$|V_{AN1}|_{rms} = \frac{|V_{AB1}|_{rms}}{\sqrt{3}} = \frac{61.35}{\sqrt{3}} = 35.42 \text{ V}$$

3. The rms value of the fundamental frequency component of the load line current $|I_{A1}|_{rms}$ is determined as:

$$|I_1|_{rms} = \frac{|V_{AN1}|}{|\overline{Z}_L|} = \frac{35.42}{\sqrt{10^2 + 7^2}} = 2.91 \text{ A}$$

4. In order to plot the current and voltage waveforms, the fundamental frequency components of the line and phase voltages, as well as the load line current are:

$$v_{AB1}(t) = 61.35\sqrt{2} \sin(120\pi t + 30°) \quad V$$

$$v_{AN1}(t) = 35.42\sqrt{2} \sin(120\pi t) \quad V$$

$$i_{A1}(t) = 2.91\sqrt{2} \sin(120\pi t - 35°) \quad A$$

Figure 6.21 shows the waveforms of the line voltage and the load line current, and the phase voltage and the load line current. Figure 6.22 shows the wave-forms of the inverter output line voltage $v_{AB}(t)$ along with its fundamental frequency component, phase voltage $v_{AN}(t)$ along with its fundamental frequency component, and the load line current $i_A(t)$ along with its fundamental frequency component.

## 6.5  SUMMARY

In this chapter, signal processing concepts of sampling and reconstructing band-limited signals from their nonuniform recurrent samples are utilized to implement and test the modulation technique for $3\phi$ voltage source (VS) six-pulse inverters. The wavelet modulation technique is based on the successful construction of a nondyadic type multiresolution analysis (MRA) capable of supporting nonuniform

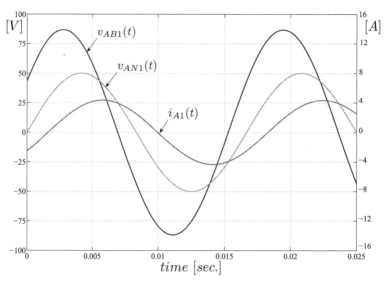

**Figure 6.21**    The 3$\phi$ Y-connected WM inverter-fed $R - L$ load: the line voltage $v_{AB1}(t)$, the phase voltage $v_{AN1}(t)$, and the load line current $i_{A1}(t)$.

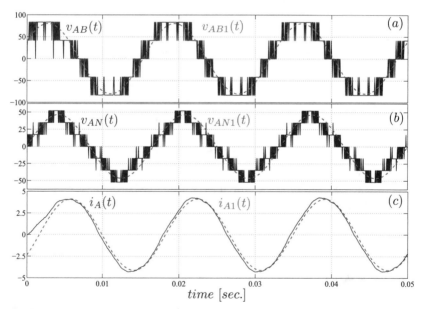

**Figure 6.22**    The 3$\phi$ Y-connected WM inverter-fed $R - L$ load: (a) the WM inverter output line voltage $v_{AB}(t)$, and its fundamental frequency component $v_{AB1}(t)$, (b) the WM inverter output phase voltage $v_{AN}(t)$ and its fundamental frequency component $v_{AN1}(t)$, and (c) the load line current $i_A(t)$, and its fundamental frequency component $i_{A1}(t)$.

recurrent sampling and reconstruction. The new scale-based linearly combined analysis and synthesis scaling functions are employed to construct the desired MRA. These scaling functions are able to generate new sets of wavelet functions capable of generating switching pulses to activate a $3\phi$ six-pulse inverter. An algorithm for implementing the wavelet modulation technique has been developed for performance testing. Simulation and experimental performance results of the $3\phi$ voltage source six-pulse WM inverter have demonstrated outputs with high fundamental component magnitudes and low harmonic contents when supplying different load types. Furthermore, performance test results show consistent features of high fundamental component magnitudes and low harmonic contents for different load types. These performance test results provide evidence for the applicability of the $3\phi$ voltage source WM inverters in various commercial and industrial applications such as renewable energy supplies, $3\phi$ ac motor drives, $3\phi$ UPS systems, power system compensation and control, and power quality applications.

# Appendix A

# Nondyadic MRA for 3φ WM Inverters

## A.1 PRELIMINARY DERIVATIONS

The scale-based linearly combined scaling function $\varphi_j(t)$ can generate basis functions $\{\varphi(2^j - k)\}_{j,k \in \mathbb{Z}}$ that constitutes Riesz bases for the multiresolution analysis (MRA) $V_j(\varphi)$. This feature of $\varphi(t)$ assures that $\varphi(t) \in L^1(\mathbb{R})$. Moreover, any periodic continuous function $f(t) \in V_0(\varphi)$ can be represented as:

$$f(t) = \sum_{n=-\infty}^{\infty} c_n \varphi(t - k) \tag{A.1}$$

where $\{c_n\}_{n \in \mathbb{Z}}$ is a unique sequence that satisfies the following:

$$\{c_n\}_{n \in \mathbb{Z}} \in l^2(\mathbb{Z}) \tag{A.2}$$

For the case of the 3φ inverter, the sampling-based model views the generation of the switching pulses as a nonuniform recurrent sampling and reconstruction process. This nonuniform recurrent sampling and reconstruction process includes sampling three sinusoidal reference modulating signals having the same frequency and amplitude with a phase shift of $2\pi/3$ *rad* from each other as:

$$S_{Ma}(t) = \sin(\omega_m t) \tag{A.3}$$

$$S_{Mb}(t) = \sin\left(\omega_m t - \frac{2\pi}{3}\right) \tag{A.4}$$

$$S_{Mc}(t) = \sin\left(\omega_m t + \frac{2\pi}{3}\right) \tag{A.5}$$

Let $\mathcal{D}$ groups of nonuniform samples be created over $T_m$. For the signal $S_{Ma}(t)$, the two samples of group $d$ will be located at:

$$t_{d1} = d + 2^{-j-1} \tag{A.6}$$

$$t_{d2} = d + 1 - 2^{-j-1} \tag{A.7}$$

*An Introduction to Wavelet Modulated Inverters*, by S.A. Saleh and M. Azizur Rahman
Copyright © 2011 Institute of Electrical and Electronics Engineers

where $d = 0, 1, 2,...\mathcal{D} - 1$, $\mathcal{D}$ is the number of created sample groups over $T_m$ and $j = 1, 2,...$ For the signal $S_{Mb}(t)$ with a phase shift of $-\dfrac{2\pi}{3}$ rad from $S_{Ma}(t)$, the two samples of group $d$ will be located at:

$$t_{d1} = d + 2^{-j-1} - \frac{T_m}{3} \tag{A.8}$$

$$t_{d2} = d + 1 - 2^{-j-1} - \frac{T_m}{3} \tag{A.9}$$

For the signal $S_{Mc}(t)$ with a phase shift of $\dfrac{2\pi}{3}$ rad from $S_{Ma}(t)$, the two samples of group $d$ will be located at:

$$t_{d1} = d + 2^{-j-1} + \frac{T_m}{3} \tag{A.10}$$

$$t_{d2} = d + 1 - 2^{-j-1} + \frac{T_m}{3} \tag{A.11}$$

where $d = 0, 1, 2,...$ $\mathcal{D} - 1$ and $j = 1, 2,...$. Equations (A.6), (A.7), (A.8), (A.9), (A.10), and (A.11) indicate that the sample group $d$ is located in different times for each sampled signal. These shifts in time locations of the sample group $d$ created at scale $j$ from $S_{Ma}(t)$, $S_{Mb}(t)$, and $S_{Mc}(t)$ indicate that simultaneous sample groups created from the three sampled signals by $\varphi_j(t)$ are from different scales.

## A.2   TIME AND SCALE LOCALIZATION OF MRA SPACES

The shifts of $\dfrac{-2\pi}{3}$ and $\dfrac{2\pi}{3}$ in $S_{Mb}(t)$ and $S_{Mc}(t)$, respectively, cause the locations of the sample group $d$ to be shifted. Evaluating sample locations $t_{d1}$ and $t_{d2}$ for $d = 0$ and $j = 1$ taken from $S_{Mb}(t)$ produces identical times to $t_{d1}$ and $t_{d2}$ evaluated for $j = 3$. Similarly, sample locations $t_{d1}$ and $t_{d2}$ for $d = 0$ and $j = 1$ taken from $S_{Mc}(t)$ produces identical times to $t_{d1}$ and $t_{d2}$ evaluated for $j = 3$. It is noted that for $S_{Mb}(t)$ and $S_{Mc}(t)$ the shifted values of the scale $j$ are identical due to the fact that the sampled signals have quarter cycle symmetry, and sample groups have symmetrical distribution around $2\pi f_m t = \dfrac{\pi}{2}$. The aforementioned discussion indicates that the simultaneous sample groups from the three signals $S_{Ma}(t)$, $S_{Mb}(t)$, and $S_{Mc}(t)$ are created with $\varphi_j(t)$ at different scales and different translations. These different scales and translations can described by shifted versions of $\varphi_j(t)$ as:

$$\left(\varphi(t)\right)_j = \phi\left(2^{j+1}t\right) + \phi\left(2^{j+1}\left(t - 1 - 2^{-(j+1)}\right)\right) \tag{A.12}$$

$$\left(\varphi(t)\right)_{jb} = \phi\left(2^{jb+1}t\right) + \phi\left(2^{jb+1}\left(t - 1 - 2^{-(jb+1)}\right)\right) \tag{A.13}$$

$$\left(\varphi(t)\right)_{jc} = \phi\left(2^{jc+1}t\right) + \phi\left(2^{jc+1}\left(t - 1 - 2^{-(jc+1)}\right)\right) \tag{A.14}$$

where $j = 1, 2...jb = 3, 4...$, and $jb = 3, 4....$ The constructed nondyadic MRA by $\{(\varphi(t))_{(j,k)}\}$, $\{(\varphi(t))_{(jb,k)}\}$, and $\{(\varphi(t))_{(jc,k)}\}$ can be formulated as:

$$V_{j,jb,jc}(\varphi) = \text{clos}_{L^2}[\langle\{\varphi_{j,k}(t)\}\rangle, \langle\{\varphi_{jb,k}(t)\}\rangle, \langle\{\varphi_{jc,k}(t)\}\rangle], \quad k \in \mathbb{Z} \qquad (A.15)$$

The nondyadic MRA formulated in equation (A.15) ensures that sample group $d$ is created simultaneously from the three signals $S_{Ma}(t)$, $S_{Mb}(t)$, and $S_{Mc}(t)$ using three sets of scaling functions from different scales. This MRA structure can be viewed as localizing in time the nested spaces of nondyadic MRA as:

$$\dots V_{(j-1),(jb-1),(jc-1)} \subset V_{(j,jb,jc)} \subset V_{(j+1),(jb+1),(jc+1)} \dots \subset L^2(\mathbb{R}) \qquad (A.16)$$

In general, the spaces of an MRA are localized in time and scale to ensure stable and accurate structure. In the case of the nondyadic MRA in equation (A.15), the scale localization condition can be satisfied through defining a time shift quantity that is capable of compensating the difference in values of $jb$ and $jc$. These time shifts can be determined as:

- For $S_{Mb}(t)$:

$$t_{d1} = d + 2^{-j-1} + z_b(j) = d_b + 2^{-j-2} \qquad (A.17)$$

$$t_{d2} = d + 1 - 2^{-j-1} + z_b(j) = d_b + 1 - 2^{-j-2} \qquad (A.18)$$

$$d_b = d + k_b(j), \quad k_b(j) \in \mathbb{R} \qquad (A.19)$$

where $k_b(j)$ is the shift in the location of sample group $d$ due to the shifted value of $jb$. Solving equations (A.17), (A.18), and (A.19) yields:

$$z_b(j) = \frac{2^{-j-2}}{3} \quad \text{and} \quad k_b(j) = \frac{2^{-j-1}}{3}$$

- For $S_{Mc}(t)$:

$$t_{d1} = d + 2^{-j-1} + z_c(j) = d_c + 2^{-j-2} \qquad (A.20)$$

$$t_{d2} = d + 1 - 2^{-j-1} + z_c(j) = d_c + 1 - 2^{-j-2} \qquad (A.21)$$

$$d_c = d + k_c(j), \quad k_c(j) \in \mathbb{R} \qquad (A.22)$$

where $k_c(j)$ is the shift in the location of sample group $d$ due to the shifted value of $jc$. The solution of equations (A.20), (A.21), and (A.22) provides:

$$z_c(j) = \frac{2^{-j-3}}{3} \quad \text{and} \quad k_c(j) = 5\frac{2^{-j-1}}{12}$$

The determination of the time shifts $z_b$ and $z_c$ provides a tool to define the shifted scaling functions that are capable to construct the nondyadic MRA, which will have spaces localized both in time and scale. These scaling functions are expressed as:

$$(\varphi_a(t))_j = \phi(2^{j+1}t) + \phi(2^{j+1}(t-u)) \qquad (A.23)$$

$$(\varphi_b(t))_j = \phi(2^{j+1}t - z_b) + \phi(2^{j+1}(t-u) - z_b) \qquad (A.24)$$

$$\left(\varphi_c(t)\right)_j = \phi\left(2^{j+1}t - z_c\right) + \phi\left(2^{j+1}(t-u) - z_c\right) \tag{A.25}$$

In the same manner, the synthesis scaling functions dual to the three shifted scaling functions can be defined as:

$$\left(\tilde{\varphi}_a(t)\right)_j = \phi_H(t) - \left(\varphi_a(t)\right)_j \tag{A.26}$$

$$\left(\tilde{\varphi}_b(t)\right)_j = \phi_H(t - z_b) - \left(\varphi_b(t)\right)_j \tag{A.27}$$

$$\left(\tilde{\varphi}_c(t)\right)_j = \phi_H(t - z_c) - \left(\varphi_c(t)\right)_j \tag{A.28}$$

where $j = 1, 2,\ldots$ The nondyadic MRA can be formulated as:

$$V_j(\varphi) = \operatorname{clos}_{L^2}\left[\left\langle \{\varphi_a(t)\}_{j,k}\right\rangle, \left\langle \{\varphi_b(t)\}_{j,k}\right\rangle, \left\langle \{\varphi_c(t)\}_{j,k}\right\rangle\right], \quad k \in \mathbb{Z} \tag{A.29}$$

Moreover, the nested spaces of the MRA in equation (A.29) can be described as:

$$\ldots V_{(j-1)} \subset V_{(j)} \subset V_{(j+1)} \ldots \subset L^2(\mathbb{R}) \tag{A.30}$$

# Bibliography

[1] Dewan, S. B. and Straughen, A., *Power Semiconductor Circuits*. John Wiley & Sons, New York, 1975.

[2] Wood, P., *Switching Power Converters*. Van Nostrand Reinhold, New York, 1981.

[3] Rashid, M. H., *Power Electronics, Circuits, Devices, and Applications*. Prentice Hall, Englewood Cliffs, NJ, 1988.

[4] Mohan, N., Undeland, T. M. and Robbins, W. P., *Power Electronics: Converters, Applications and Design*, John Wiley & Sons, New York, 1989.

[5] Chui, C. K., *An Introduction to Wavelets*, $3^{rd}$ Edition. Academic Press, San Diego, CA, 1992.

[6] Daubechies, I., *Ten Lectures on Wavelets*, $3^{rd}$ Edition. SIAM: Society of Industrial and Applied Mathematics Series, Philadelphia, PA, 1992.

[7] Strang, G. and Nguyen, T., *Wavelets and Wavelet Filter Banks*. Wellesly-Cambridge Press, Wellesley, MA, 1996.

[8] Chui, C. K., *Wavelets: A Mathematical Tool for Signal Processing*. SIAM: Society of Industrial and Applied Mathematics Series, Philadelphia, PA, 1997.

[9] Oppenhiem, A. V., Schafer, R. W. and Buck, J. R., *Discrete Time Signal Processing*, $2^{nd}$ Edition, Prentice Hall, Upper Saddle River, NJ, 1999.

[10] Bose, B. K., *Modern Power Electronics and AC Drives*, Prentice Hall, Upper Saddle River, NJ, 2002.

[11] Holmes, D. G. and Lipo, T. A., *Pulse Width Modulation for Power Converters*. IEEE Press Series on Power Engineering, Wiley Inter-Science, Hoboken NJ, 2003.

[12] Schaumann, R. and Van Valkenburg, M. E., *Design of Analog Filters*, $4^{th}$ Edition. Oxford University Press, New York, 2005.

[13] *Power System Toolbox User Guide*. Math Works, Natick, MA, 2007.

[14] Digital Signal Processing and Control Engineering, *dSPACE Gmbh*, Paderborn, Germany, 2007.

[15] Kotsakis, C., Multiresolution Aspects of Linear Approximation Methods in Hilbert Spaces Using Gridded Data. Ph.D. thesis, University of Calgary, Calgary, AL, Canada, 2000.

[16] Saleh, S. A., Wavelet Modulated DC-AC Power Inverters. Ph.D. thesis, Memorial University of Newfoundland, St. John's, NF, Canada, 2007.

[17] Paley, R. and Wiener, N., "Fourier Transform in the Complex Domain," *Amer. Math. Soc. Colloq. Publications*, Vol. 19, 1934.

[18] Levinson, N., "Gap and Density Theorems," *Amer. Math. Soc. Colloq. Publications*, Vol. 26, 1940.

[19] Duffin, R. and Schaeffer, A., "A Class of Nonharmonic Fourier Series," *Trans. Amer. Math. Soc.*, Vol. 72, pp. 341–366, 1952.

[20] Royer, G. H., "A Switching Transistor D-C to A-C Converter Having an Output Frequency Proportional to the D-C Input Voltage," *AIEE Trans. on Communications and Electronics*, Vol. 74, Part. I, pp. 322–324, 1955.

[21] Uchrin, G. C. and Taylor, W. O., "A New Self-Excited Square-Wave Transistor Power Oscillator," *Proc. of the IRE*, Vol. 43, pp. 99, 1955.

[22] Kirnick, A. and Heinrick, "Static Inverters with Neutralization of Harmonics," *AIEE Transactions*, Vol. 81, pp. 374–378, 1962.

[23] Schonubg, A. and Stemmler, H., "Static Frequency Changers with Subharmonic Control in Conjunction with Reversible Variable Speed AC Drives," *Brown Boveri Rev.*, pp. 555–557, 1964.

[24] Mokrytzki, B., "Pulse Width Modulated Inverters for AC Motor Drives," *IEEE Trans. on IA*, Vol.-IGA-3, pp. 493–503, 1967.

[25] McMurry, W., "Multipurpose Power Converter Circuits," *U.S. Patent No.* 3,487,289, December, 1969.

[26] McMurry, W., "Analysis of Thyristor DC Chopper Power Converters Including Nonlinear Commutating Reactors," *IEEE Trans. on Magnetics*, Vol. MAG-1, pp. 16–21, 1970.

[27] Patel, H. S. and Hoft, R. G., "Generalised Techniques of Harmonic Elimination and Voltage Control in Thyristor Inverters: Part I Harmonic Elimination," *IEEE Trans. on Industry Applications*, IA-9(3), pp. 310–317, 1973.

[28] Lee, Y. F. and Willson, T. G., "Analysis and Modeling of a Family of TwoTransistor Parallel Inverters," *IEEE Trans. on Magnetics*, Vol. Mag-9, No. 3, pp. 414–418, 1973.

[29] Patel, H. S. and Hoft, R. G., "Generalised Techniques of Harmonic Elimination and Voltage Control in Thyristor Inverters: Part II - Voltage Control Techniques," *IEEE Trans. on Industry Applications*, IA-10(5), pp. 666–673, 1974.

[30] Bowes, R. S. and Bird, B. M., "Novel Approach to the Analysis and Synthesis of Modulation Processes in Power Converters," *Proc. IEE.*, Vol. 122, No. 5, pp. 507–513, 1975.

[31] Bowes, S. R., "New Sinusoidal Pulsewidth-Modulated Inverter," *Proc. IEE.*, Vol. 122, No. 11, pp. 1279–1285, 1975.

[32] Papoulis, A., "Generalized Sampling Expansion," *IEEE Trans. on Circuits and Systems*, Vol. 24, No. 11, pp. 652–654, 1977.

[33] Buja, G. and Indri, G., "Optimal PWM for Feeding AC Motors," *IEEE Trans. on Industry Applications*, Vol. 13, No. 1, pp. 34–42, 1977.

[34] Plunckett, A. B., "A Current Controlled PWM Transistor Inverter Drive," *IEEE/IAS 1979 Annual Meeting Proc.*, pp. 785–792, 1979.

[35] Grant, D. A. and Seinder, R., "Ratio Changing in Pulse Width Modulated Inverters," *Proc. IEE*, Vol. 128, Part B, No. 5, pp. 243–248, 1981.

[36] Bowes, S. R. and Mount, M. J., "Microprocessor Control of PWM Inverters," *IEE Proc., B, Elect. Power Appl*, Vol. 128, No. 6, pp. 293–305, 1981.

[37] Nabae, A., Takahashi, I. and Akagi, H., "A New Neutral-Point-Clamped PWM Inverter," *IEEE Trans. on Industry Applications*, Vol. 17, No. 5, pp. 518–523, 1981.

[38] Ziogas, P. D., "The Delta Modulation Technique in Static PWM Inverters," *IEEE Trans. on Industry Applications*, Vol. 17, No. 2, pp. 199–204, 1981.

[39] Holtz, J. and Stadtfeld, S., "A Predictive Controller for the Stator Current Vector of AC Machines Fed from a Switched Voltage Source," *IPEC'83 Conference Record*, Tokyo, pp. 1165–1675, 1983.

[40] Houndsworth, J. A. and Grant, D. A., "The Use of Harmonic Distortion to Increase Voltage of a Three Phase PWM Inverter," *IEEE Trans. on Industry Applications*, Vol. IA-20, pp. 1224–1228, 1984.

[41] Rowan, T. R. and Kerkman, R. L., "A New Synchronous Current Regulator and an Analysis of Current-Regulated PWM Inverters," *IEEE Trans. on Industry Applications*, Vol. 22, No. 4, pp. 678–690, 1986.

[42] Enjeti, P. and Lindsay, J. F., "Solving Nonlinear Equations of Harmonic Elimination PWM in Power Control," *IEE Electronics Letters*, Vol. 23, pp. 656–657, 1987.

[43] Trzynadlowski, A. M., Legowaki, S. and Kirlin, R. L., "Random Pulse Width Modulation Technique for Voltage-Controlled Power Inverters," *IEEE/IAS 1987 Annual Meeting Proc.*, pp. 863–868, 1987.

[44] Rahman, M. A., Quaicoe, J. E. and Choudhury, M. A., "Performance Analysis of Delta Modulated Inverters," *IEEE Trans. on Power Electronics*, Vol. 2, No. 3, pp. 227–233, 1987.

[45] Boost, M. A. and Ziogas, P. D., "State-of-the-art Carrier PWM Techniques: a Critical Evaluation," *IEEE Trans. on Industry Applications*, Vol. 24, No. 2, pp. 271–280, 1988.

[46] Daubechies, I., "Orthonormal Bases of Compactly Supported Wavelets," *Comm. Pure & Appl. Math.*, Vol. 41, pp. 909–996, 1988.

[47] Trzynadlowski, A. M., "Nonsinusoidal Modulating Functions for Three-Phase Inverters," *IEEE Trans. on Power Electronics*, Vol. 4, No. 3, pp. 331–338, 1989.

[48] Mallat, S. G., "A Theory for Multiresolution Signal Decomposition: the Wavelet Representation," *IEEE Trans. on Pattern Analysis and Machine Intelligence*, Vol. 11, No. 7, pp. 674–693, 1989.

[49] Strang, G., "Wavelets and Dilation Equations: A Brief Introduction," *SIAM Rev.*, Vol. 31, pp. 614–627, 1989.

[50] Berry, F., "Steady State Mathematical Model for the DC-AC Inverters on the Space Shuttle," *IEEE Conference on Energy and Information Technologies in the Southeast*, Southeast Con'89 Proceedings, Vol. 2, pp. 455–458, 1989.

[51] Enjeti, N., Ziogas, P. D. and Lindsay, J. F., "Programmed PWM Techniques to Eliminate Harmonics: a Critical Evaluation," *IEEE Trans. on Industry Applications*, Vol. 26, No. 2, pp. 302–316, 1990.

[52] Bose, B. K., "An Adaptive Hysteresis-Band Current Control Technique of a Voltage-Fed PWM Inverter for Machine Drive System," *IEEE Trans. on Industrial Electronics*, Vol. 37, No. 5, pp. 402–408, 1990.

[53] Zayed, A. I., "On Kramer's Sampling Theorem Associated with General Sturm-Liouville Problems and Lagrange Interpolation," *SIAM Journal Applied Math.*, Vol. 51, pp. 575–604, 1991.

[54] Tymerski, R. E., "Frequency Analysis of Time-Interval-Modulated Switched Networks," *IEEE Trans. on Power Electronics*, Vol. 6, No. 2, pp. 287–295, 1991.

[55] Nashed, M. Z. and Walter, G. G., "General Sampling Theorems for Functions in Reproducing Kernel Hilbert Spaces," *Math. Control Signals Sys.*, Vol. 4, pp. 373–412, 1991.

[56] Bowes, S. R. and Clark, P. R., "Simple Microprocessor Implementation of New Regular-Sampled Harmonic Elimination PWM Techniques," *IEEE Trans. on Industry Applications*, Vol. 28, No. 1, pp. 89–95, 1992.

[57] Walter, G. G., "A Sampling Theorem for Wavelet Subspaces," *IEEE Trans. on Information Theory*, Vol. 38, No. 2, pp. 881–884, 1992.

[58] Odegard, J.E., Gopinath, R.A. and Burrus, C.S., "Optimal Wavelets for Signal Decomposition and the Existence of Scale-Limited Signals," *IEEE International Conference on Acoustics, Speech, and Signal Processing, ICASSP-92 Record*, Philadelphia, PA, Vol. 4, pp. 597–600, 1992.

[59] Unser, M., Aldroubi, A. and Eden, M., "On the Asymptotic Convergence of B-spline Wavelets to Gabor Functions," *IEEE Trans. on Information Theory*, Vol. 38, pp. 864–872, March 1992.

[60] Bowes, S. R. and Clark, P. R., "Transputer-Based Optimal PWM Control of Inverter Drives," *IEEE Trans. on Industry Applications*, Vol. 28, No. 1 Part 1, pp. 81–88, 1992.

[61] Vetterli, M. and Herley, C., "Wavelets and Filter Banks: Theory and Design," *IEEE Trans. on Signal Processing*, Vol. 40, No. 9, pp. 2207–2232, 1992.

[62] Gilbert, S., "Wavelet Transforms Versus Fourier Transforms," *Bull. Amer. Math. Soc.*, Vol. 28, pp. 288–305, 1993.

[63] Xia, X. G. and Zhang, Z., "On Sampling Theorem, Wavelets, and Wavelet Transforms," *IEEE Trans. on Signal Processing*, Vol. 41, No. 12, pp. 3524–3535, 1993.

[64] Abdel-Rahim, N. and Quaicoe, J. E., "A Single-Phase Delta-Modulated Inverter for UPS Applications," *IEEE Trans. on Industrial Electronics*, Vol. 40, No. 3, pp. 347–354, 1993.

[65] Gopinath, R. A., Odegard, J. E. and Burrus, C. S., "Optimal Wavelet Representation of Signals and the Wavelet Sampling Theorem," *IEEE Trans. on Circuits and Systems-II: Analog and Digital Signal Processing*, Vol. 41, No. 4, pp. 262–277, 1994.

[66] Bowes, S. R., "Novel Real-Time Harmonic Minimized PWM Control for Drives and Static Power Converters," *IEEE Trans. on Power Electronics*, Vol. 9, No. 3, pp. 256–262, 1994.

[67] Bowes, S. R., "Advanced Regular-Sampled PWM Control Techniques for Drives and Static Power Converters," *IEEE Trans. on Industrial Electronics*, Vol. 42, No. 4, pp. 367–373, 1995.

[68] Abdel-Rahim, N. and Quaicoe, J. E., "Analysis and Design of a Multiple Feed-back Loop Control Strategy for Single-Phase Voltage-Source UPS Inverters," *IEEE Trans. on Power Electronics*, Vol. 11, No. 4, pp. 532–541, 1996.

[69] Bowes, R. S. and Lai, Y. S., "The Relationship Between Space-Vector Modulation and Regular-Sampled PWM," *IEEE Trans. on Industrial Electronics*, Vol. 44, No. 5, pp. 670–679, 1997.

[70] Rahman, K. M., Rezwan, M. K., Choudhury, M. A. and Rahman, M. A., "Variable-Band Hysteresis Current Controllers for PWM Voltage-Source Inverters," *IEEE Trans. on Power Electronics*, Vol. 12, No. 6, pp. 964–970, 1997.

**138**    Bibliography

[71] Liang, T. J., O'Connell, R. M. and Hoft, R.G., "Inverter Harmonic Reduction Using Walsh Function Harmonic Elimination Method," *IEEE Trans. on Power Electronics*, Vol. 12, No. 6, pp. 971–982, 1997.

[72] Hui, S. R., Sathiakumar, S. and Sung, K. K., "Novel Random PWM Scheme with Weighted Switching Decision," *IEEE Trans. on Power Electronics*, Vol. 12, No. 6, pp. 945–952, 1997.

[73] Stankovic, A. M., Verghese, G. C. and Perreault, D. J., "Randomized Modulation of Power Converters Via Markov Chains," *IEEE Trans. on Control System Technology*, Vol. 5, No. 1, pp. 61–73, 1997.

[74] Pandy, S. K. and Satish, L., "Multiresolution Signal Decomposition: New Tool for Fault Detection in Power Transformers During Impulse Tests," *IEEE Trans. on Power Delivery*, Vol. 13, No. 4, pp. 1194–1200, 1998.

[75] Unser, M. and Zerubia, J., "A Generalized Sampling Theory without Bandlimiting Constraints," *IEEE Trans. on Circuits and Systems II*, Vol. 45, No. 8, pp. 959–969, 1998.

[76] Unser, M., Thevenaz, P. and Aldroubi, A., "Shift-Orthogonal Wavelet Bases," *IEEE Trans. on Signal Processing*, Vol. 46, No. 7, pp. 1827–1836, 1998.

[77] Bölcskei, H., "Oversampling in Wavelet Subspaces," *Proc. IEEE-SP 1998 Int. Sympos. Time-Frequency Time-Scale Analysis*, Pittsburgh, PA, pp. 489–492, Oct. 1998.

[78] Selesnick, I. W., "Interpolating Multiwavelet Bases and the Sampling Theorem," *IEEE Trans. on Signal Processing*, Vol. 47, No. 6, pp. 1615–1621, 1999.

[79] Walker, G. and Ledwich, G., "Bandwidth Considerations for Multilevel Converters," *IEEE Trans. on Power Electronics*, Vol. 14, No. 1, pp. 74–81, 1999.

[80] Jia, R. Q., Jiang, Q. and Shen, Z., "Distributional Solutions of Non-Homogenous Discrete and Continuous Refinement Equations," *SIAM Journal on Applied Math.*, Vol. 32, No. 2, pp. 420–434, 2000.

[81] Bowes, S. R. and Jian L., "New Robust Adaptive Control Algorithm for High-Performance AC Drives," *IEEE Trans. on Industrial Electronics*, Vol. 47, No. 2, pp. 325–336, 2000.

[82] Eldar C. Y. and Oppenheim, A. V., "Filterbank Reconstruction of Bandlimited Signals from Nonuniform and Generalized Samples," *IEEE Trans. on Signal Processing*, Vol. 18, No. 10, pp. 2864–2875, 2000.

[83] Celanovic, N. and Boroyevich, D., "A Comprehensive Study of Neutral-Point Voltage Balancing Problem in Three-Level Neutral-Point-Clamped Voltage Source PWM Inverters," *IEEE Trans. on Power Electronics*, Vol. 15, No. 2, pp. 242–249, 2000.

[84] H. Zhang, H., Jouanne, A. V., Dai, S., Wallace, A. K. and Wang, F., "Inverter Modulation Schemes to Eliminate Common-Mode Voltages," *IEEE Trans. on Industry Applications*, Vol. 36, No. 6, pp. 1645–1653, 2000.

[85] Zhang, X., Muguruma, T. and Yoshikawa, T., "Design of Orthonormal Symmetric Wavelet Filters Using Real Allpass Filters", *Signal Processing*, Vol. 80, No. 8, pp. 1551–1559, August 2000.

[86] Maswood, A. I., Shen, W. and Rahman, M. A., "A Flexible Way to Generate PWM-SHE Switching Patterns Using Genetic Algorithm," *IEEE APEC 2001*, Vol. 2, pp. 1130–1134, 2001.

[87] Raman, V. and Yoram, B., "Optimal Sub-Nyquist Nonuniform Sampling and Reconstruction for Multiband Signals," *IEEE Trans. on Signal Processing*, Vol. 49, No. 10, pp. 2301–2313, 2001.

[88] Aldroubi, A. and Grochenig, K., "Non-uniform Sampling and Reconstruction in Shift-Invariant Spaces," *SIAM Rev.*, Vol. 43, No. 4, pp. 585–620, 2001.

[89] Koutroulis, E., Chatzakis, J., Kalaitzakis, K. and Voulgaris, N. C., "A Bidirectional, Sinusoidal, High-Frequency Inverter Design," *IEE Proc.-Electr. Power Appl.*, Vol. 148, No. 4, pp. 315–321, 2001.

[90] Dogaru, T. and Carin, L., "Multiresolution Time-Domain Using CDF Biorthogonal Wavelets," *IEEE Trans. on Microwave Theory and Techniques*, Vol. 49, No. 5, pp. 902–912, 2001.

[91] Zeng, F. P., "A Generalized Multilevel Inverter Topology with Self Voltage Balancing," *IEEE Trans. on Industry Applications*, Vol. 37, No. 2, pp. 611–618, 2001.

[92] Dogaru, T. and Carin, L., "Multiresolution Time-Domain Using CDF Biorthogonal Wavelets," *IEEE Trans. on Microwave Theory and Techniques*, Vol. 49, No. 5, pp. 902–912, 2001.

[93] Czarkowski, D., Chudnovsky, D. V., Chudnovsky, G. V. and Selesnick, I. W., "Solving the Optimal PWM Problem for Single-Phase Inverters," *IEEE Trans. on Circuits and Systems -I: Fundamental Theory and Applications*, Vol. 49, No. 4, pp. 465–475, 2002.

[94] Yamanaka, K., Hava, A. H., Kirino, H., Tanaka, Y., Koga, N. and Kume, K., "A Novel Neutral Point Potential Stabilization Technique Using the Information of Output Current Polarities and Voltage Vector," *IEEE Trans. on Industry Applications*, Vol. 38, No. 6, pp. 1572–1580, 2002.

[95] Rodriguez, J., Lai, J. S. and Zeng, F. P., "A Survey of Topologies, Controls and Applications," *IEEE Trans. on Industrial Electronics*, Vol. 49, No. 4, pp. 724–7389, 2002.

[96] Hamid, E. Y. and Kawasaki, Z. I., "Wavelet-Based Data Compression for Power Disturbances Using Minimum Description Length Data," *IEEE Trans. on Power Delivery*, Vol. 17, No. 2, pp. 460–466, 2002.

[97] Rech, C., Pinheiro, H., Grundling, H. A., Hey, H. L. and Pinheiro, J. R., "A Modified Discrete Control Law for UPS Applications," *IEEE Trans. on Power Electronics*, Vol. 18, No. 5, pp. 1138–1145, 2003.

[98] Zhang, K., Kang, Y., Xiong, J. and Chen, J., "Direct Repetitive Control of SPWM Inverter for UPS Purpose," *IEEE Trans. on Power Electronics*, Vol. 18, No. 3, pp. 784–792, 2003.

[99] Zhang, J. K., Davidson, T. N. and Wong, K. M., "Efficient Design of Orthonormal Wavelet Bases for Signal Representation," *IEEE Trans. on Signal Processing*, Vol. 52, No. 7, pp. 1983–1996, 2004.

[100] Milosevic, M., "Decoupling Control of $d$ and $q$ Current Components in Three-Phase Voltage Source Inverter," *EEH Power Systems Laboratory, Technical Report*, Zuerich, Switzerland, 2004.

[101] Saleh, S. A. and Rahman, M. A., "Discrete Time-Based Model of the Sinusoidal Pulse Width Modulation Technique," *IEEE IECON'05 Conference Proceeding*, Raleigh, NC., pp. 1082–1087, November 2005.

[102] Van De Ville, D., Blu, T., Forster, B. and Unser, M.,"Semi-Orthogonal Wavelets that Behave Like Fractional Differentiators," *Proceedings of the SPIE: Optical Engineering and Instrumentation (Wavelet XI)*, San Diego, CA, August, 2005.

[103] Tam, K. C., Wong, S. C. and Tse, C. K., "Wavelet-Based Piecewise Approximation of Steady-State Waveforms for Power Electronics Circuits," *The IEEE International Symposium on Circuits and Systems (ISCAS'05)*, Kobe, Japan, Vol. 3, pp. 2490–2493, May 2005.

[104] Chang, W. G. and Chen, S. K., "An Analytical Approach for Characterizing Harmonic and Inter-harmonic Currents Generated by VSI-Fed Adjustable Speed Drives," *IEEE Trans. on Power Delivery*, Vol. 20, No. 4, pp. 2585–2593, 2005.

[105] Guinee, R. A. and Lyden, C., "A Novel Fourier Series Time Function for Modeling and Simulation of PWM," *IEEE Trans. on Circuits and Systems–I: Regular Papers*, Vol. 52, No. 11, pp. 2427–2435, 2005.

[106] Saleh, S. A. and Rahman, M. A., "A New Transient Model for Three-Phase Power Transformers Using a Wavelet Filter Bank," *IEEE Trans. on Power Delivery*, Vol. 20, No. 2, pp. 1409–1416, 2005.

[107] Meco-Gutierrez, M. J., Perez-Hidalgo, F., Vargas-Merino, F. and Heredia Larrubia, J. R., "A New PWM Technique Frequency Regulated Carrier for Induction Motor Supply," *IEEE Trans. on Industrial Electronics*, Vol. 53, No. 5, pp. 1750–1754, 2006.

[108] Bowes, S. R. and Holliday, D., "Comparison of Pulse-Width-Modulation Control Strategies for Three-Phase Inverter Systems," *IEE Proceedings: Electric Power Applications*, Vol. 153, No. 4, pp. 575–584, 2006.

[109] Hadiouche, D., Baghli, L. and Rezzoug, A., "Space-Vector PWM Techniques for Dual Three-Phase AC Machine: Analysis, Performance Evaluation, and DSP Implementation," *IEEE Trans. on Industry Applications*, Vol. 42, No. 4, pp. 1112–1122, 2006.

[110] Saleh, S. A., Moloney, C., R. and Rahman, M. A., "Developing a Non-dyadic MRAS for Switching DC-AC Inverters," *The IEEE 12th Digital Signal Processing Workshop DSP06 Conference*, Jackson Lake Lodge, WY, September, pp. 544–549, 2006.

[111] Saleh, S. A. and Rahman, M. A., "Development and Experimental Testing of a Single-Phase *B*-Spline-Based SPWM Inverter," *IEEE ISIE'06 Conference Proceedings*, Montreal, Quebec, pp. 815–819, July, 2006.

[112] Saleh, S. A. and Rahman, M. A., "Experimental Testing of a Novel Control for Inverter-Fed Three-Phase Induction Motor," *IEEE PES'06 Conference Proceedings*, Montreal, Quebec, June, 2006.

[113] Koizumi, H., Kurokawa, K. and Mori, S., "Analysis of Class D Inverter With Irregular Driving Patterns," *IEEE Trans. on Circuits and Systems–I: Regular Papers*, Vol. 53, No. 3, pp. 677–687, 2006.

[114] Daubechies, I., DeVore, R. A., Güntürk, C. S. and Vaishampayan, V. A., "A/D Conversion With Imperfect Quantizers," *IEEE Trans. on Information Theory*, Vol. 52, No. 3, pp. 874–885, 2006.

[115] Donoho, D. L., "Compressed Sensing," *IEEE Trans. on Information Theory*, Vol. 52, No. 4, pp. 1289–1306, 2006.

[116] Salmon, J., Wang, L., Noor, N. and Krieger, A. W., "A Carrier-Based Unipolar PWM Current Controller That Minimizes the PWM-Cycle Average Current-Error Using Internal Feedback of the PWM Signals," *IEEE Trans. on Power Electronics*, Vol. 22, No. 5, pp. 1708–1718, 2007.

[117] Cho, K. M., Oh, W. S., Kim, Y. T. and Kim, H. J., "A New Switching Strategy for Pulse Width Modulation (PWM) Power Converters," *IEEE Trans. on Industrial Electronics*, Vol. 54, No. 1, pp. 330–337, 2007.

[118] Sundareswaran, K., Jayant, K. and Shanavas, T. N., "Inverter Harmonic Elimination Through a Colony of Continuously Exploring Ants," *IEEE Trans. on Industrial Electronics*, Vol. 54, No. 5, pp. 2558–2565, 2007.

[119] Shu, Z., Tang, J., Guo, Y. and Lian, J., "An Efficient SVPWM Algorithm With Low Computational Overhead for Three-Phase Inverters," *IEEE Trans. on Power Electronics*, Vol. 22, No. 5, pp. 1797–1805, 2007.

[120] Serpa, L. A., Round, S. D. and Kolar, J. W., "A Virtual-Flux Decoupling Hysteresis Current Controller for Mains Connected Inverter Systems," *IEEE Trans. on Power Electronics*, Vol. 22, No. 5, pp. 1766–1777, 2007.

[121] Saleh, S. A., Radwan, T. S. and Rahman, M. A., "Real-Time Testing of WPT Based Protection of Three-Phase VS PWM Inverter-Fed Motors," *IEEE Trans. on Power Delivery*, Vol. 22, No. 4, pp. 2108–2115, 2007.

[122] Saleh, S. A. and Rahman, M. A., "Real-Time Performance Testing of a $3\phi$ VS WM Inverter-Fed IPM Motor," *IEEE PESC'07 Conference Proceeding*, Orlando, FL, pp. 437–442, June 2007.

[123] Saleh, S. A. and Rahman, M. A., "Development and Experimental Testing of a $3\phi$ Resolution-Level Controlled WM Inverter-Fed Induction Motor," *IEEE IAS Annual Meeting IAS'07 Proceedings*, New Orleans, LA, pp. 1249–1256, 2007.

[124] Saleh, S. A., Moloney, C. R. and Rahman, M. A., "Implementation of a Dynamic Voltage Restorer System Based on Discrete Wavelet Transforms," *IEEE Trans. on Power Delivery*, Vol. 23, No. 4, pp. 2366–2375, 2008.

[125] Acharya, K., Mazumder, S. K. and Basu, I., "Reaching Criterion of a Three-Phase Voltage-Source Inverter Operating with Passive and Nonlinear Loads and Its Impact on Global Stability," *IEEE Trans. on Industrial Electronics*, Vol. 55, No. 4, pp. 1795–1812, 2008.

[126] Narayanan, G., Zhao, D., Krishnamurthy, H. K., Ayyanar, R. and Ranganathan, V. T., "Space Vector Based Hybrid PWM Techniques for Reduced Current Ripple," *IEEE Trans. on Industrial Electronics*, Vol. 55, No. 4, pp. 1614–1627, 2008.

[127] Liao, L. C., Pan, C. T. and Jong, T. L., "Switching Flow-Graph Modeling Technique for Three-Phase Inverters," *IEEE Trans. on Industrial Electronics*, Vol. 55, No. 4, pp. 1603–1613, 2008.

[128] Khaligh, A. R., Wells, J. R., Chapman, P. L. and Krein, P. T., "Dead-Time Distortion in Generalized Selective Harmonic Control," *IEEE Trans. on Power Electronics*, Vol. 23, No. 3, pp. 11511–1517, 2008.

[129] Bose, B. K., "Power Electronics and Motor Drives Recent Progress and Perspective," *IEEE Trans. on Industrial Electronics*, Vol. 56, No. 2, pp. 581–588, 2009.

[130] Basu, K., Siva Prasad, J. and Narayanan, G., "Minimization of Torque Ripple in PWM AC Drives," *IEEE Trans. on Industrial Electronics*, Vol. 56, No. 2, pp. 553–558, 2009.

[131] Saleh, S. A. and Rahman, M. A., "Analysis and Real-Time Testing of a Controlled Single-Phase Wavelet-Modulated Inverter for Capacitor-Run Induction Motors," *IEEE Trans. on Energy Conversion*, Vol. 24, No. 1, pp. 21–29, 2009.
[132] Saleh, S. A., Moloney, C. R. and Rahman, M. A., "Development and Testing of Wavelet Modulation for Single-Phase Inverters," *IEEE Trans. on Industrial Electronics*, Vol. 56, No. 7, pp. 2588–2599, 2009.

# Index

AC, 1
  motor
    capacitor-run, 97
    drives, 1
    three phase squirrel-cage induction, 115
    universal, 91
  variable-frequency drives, 5
approximation, 19
  $N$-dimensional, 19
  piece-wise linear, 14

bipolar junction transistor (BJT), 1

circuit, 1
  open, 2
  short, 2
  theory, 1
circuit element
  capacitive, 2
  inductive, 2
communication, 16
  applications, 16
  systems, 18
control systems, 1, 18
current, 1
  $3\varphi$ line, 111
  hall-effect sensor, 95
  load, 90
  neutral paths, 15
  source, 2
  THD$I$, see inverter, 90, 115

DC, 1
digital systems, 12
diode, 7
diode rectifier
  single phase full-wave, 92, 98
  three phase full-wave, 117

dSPACE ds1104 board, 95, 119
dyadic MRA, see multiresolution analysis

electric motors, 15
  overheat, 15
energy, 1
  electric, 1
  renewable, 1
equation
  refinement, 69
execution time, 100
expanding matrix, 22
  determinant, 22
  eigenvalues, 22

filter
  refinement, 69, 73
filter banks, 36, 45
  quadrature mirror (QMF), 36
Fourier Series, 15
  coefficients, 15
  complex coefficients, 16
  double-frequency terms, 16
  frequency components
    magnitude, 16
    phase, 16
frequency, 1
  components, see Fourier Series
    transient, 18
  domain, 25
  even components, 15
  fundamental component, 16
  narrow band, 13
  Nyquist, 34
  time-localized scales, 18
  time-locations, 26
function, 15
  $B$-spline, 24
  basis, 20, 28

*An Introduction to Wavelet Modulated Inverters*, by S.A. Saleh and M. Azizur Rahman
Copyright © 2011 Institute of Electrical and Electronics Engineers

**143**

function (*cont'd*)
  applications, 28
  bi-orthogonal, 26
  classes, 22
  complementary, 25
  completeness, 21
  contracted high frequency, 20
  dense, 19
  dilated low frequency, 20
  dyadic dilations, 18
  Fourier, 22
  Haar, 22
  Hilbert space, 21
  integer-indexed translations, 18
  linear combination, 20
  linear independence, 21
  localization, 21
  non-orthonormalized, 25
  orthogonal, 18, 22, 23
  orthogonal complement, 24
  orthonormal, 20, 21, 23
  orthonormalization, 23
  Riesz basis, 23
  scale-based linearly-combined, 67
  scaling, 21
  semi-orthogonal, 24
  shift-orthogonal, 27
  wavelet, 21
Dirac Delta, 33
interpolating, 77
  interval of support, 71
Lagrange interpolating, 34, 35, 46, 76
  derivative, 46, 55
  interval of support, 47
linear-time variable, 16
odd, 15
periodic, 15
  period, 16
sampling, 31
scaling, 18, 20
analysis, 27, 62
bi-orthogonal, 26
conditions, 20
continuous, 20
Daubechies, 24
decaying property, 20
dilation, 76
dilation equation, 20, 22
dual, 26

fractional *B*-spline, 24, 25
Haar, 24, 63, 69
integer translations, 20
non-integer translations, 72
number of vanishing moments, 64
order balance, 70
Riesz basis, 22
scale-based linearly-combined, 66, 67,
  69, 72, 73, 76, 79–81, 86
shifted scale-based linearly-combined,
  81, 82, 109
shifted synthesis scale-based
  linearly-combined, 83
synthesis, 27, 62
synthesis scale-based
  linearly-combined, 77, 79, 86, 87
Shannon sampling, 29
wavelet, 21
  analysis, 62
  Haar, 66
  integer translations, 21
  scale-based linearly-combined, 73, 74
  synthesis, 62
  wavelet basis, 19, 20

Gram-Schmidt procedure, 23

harmonics, 1, 15
  distortion, 1, 11
  energy distribution, 15
  frequencies, 15
  generation, 1, 11
  magnitudes, 15
  multiples of three, 5

industrial applications, 1
insulated gate bipolar transistor (IGBT), 1
integration, 16
  double formula, 16
inverter, 1
  carrier signal frequency, 13
  current source (CS), 2
  DC voltage source, 9
  efficiency, 12
  function, 80
  input DC voltage, 90
  instantaneous switching actions, 40
  linear modulation range, 11, 13
  loading conditions, 11

model, 39
  circuit theory, 39
  non-uniform recurrent samples'
    groups, 42
  non-uniform recurrent
    sampling-based, 47, 75
  numerical data, 39
  small-signal analysis, 40
  steady-state, 39
  time-averaging, 39
  unified converter theory, 39
modulation techniques, 11
  advantages, 15
  algorithms, 15
  approaches, 12
  carrier signal, 12
  carrier-based PWM, 13
  delta modulation (DM), 14
  modulating signal, 13
  naturally sampled PWM, 16
  non-sinusoidal PWM, 13
  performance comparison, 100, 126
  pulse-width modulation (PWM), 12
  random PWM (RPWM), 13
  regularly sampled PWM, 16
  selected harmonic elimination (SHE),
    13
  sinusoidal PWM (SPWM), 13
  space-vector modulation (SVM), 13
  wavelet modulation (WM), 97, 122
multi-level, 6
  $1\varphi$ cascaded single-phase $H$-bridge, 9
  $3\varphi$ capacitor-clamped multi-level
    (CCM), 8
  $3\varphi$ diode-clamped multi-level (DCM), 7
  $3\varphi$ neutral-point clamped (NPC), 6
  back-to-back-diode-clamped multi-
    level, 10
  generalized multi-level, 10
  mixed-level, 10
  soft-switched multi-level, 10
  zero-voltage point, 6
multi-switching techniques, 39
  modulation process, 39
operation, 16
  requirements, 65
output, 1
  energy, 10
  frequency, 15

fundamental frequency component, 10,
  91
harmonic contents, 39, 92
harmonic elimination, 39
harmonic frequency side-bands,
  41
harmonic spectrum, 41, 91
harmonics reduction, 11
jitter, 39
line-to-line, 52
line-to-neutral, 52
magnitude spectrum, 13, 91
mathematical expression, 109
per-unit, 12
piece-wise linear, 40
quality, 10, 12
six-step waveform, 5
square-wave, 5
synthesis, 8
total harmonic distortion factor (THD),
  12, 16, 91
trains of rectangular pulses, 76
waveforms, 3
performance, 1
  comparison, 15
  evaluation, 12
  open-loop steady-state, 91
sequential switching methods, 10
single phase ($1\varphi$), 2, 88
  full-bridge ($H$-bridge), 2
  half-bridge (center–tapped), 2
  schematic diagram, 85
  voltage source IGBT $H$-bridge, 91
switch utilization, 7
switching actions, 2
switching elements, 1
  IGBT, 95
  matrix, 40
  operation, 1
  sequential activation, 7
  status ON, OFF, 2
  switch-cell, 40
  switching time, 87
switching frequency, 12
switching losses, 12
switching strategies, 11
  multi-switching, 11
  single-pulse, 13
  square-wave, 11

inverter (*cont'd*)
  three phase ($3\varphi$), 2, 49, 79
    120°-conduction, 5
    180°-conduction, 5
    3-leg six-pulse, 50
    active switches, 54
    applications, 88
    current source (CS), 52
    sampling-based model, 57, 88
    voltage source (VS), 52
    voltage source (VS) six-pulse,
      111
  topologies, 2, 3, 6
  voltage source (VS), 2
  wavelet modulation technique, 85
    experimental performance, 95, 119
    implementation, 89, 90, 109
    implementation procedure, 109
    performance simulation, 90, 111
    real time implementation, 95, 119

load, 89
  $3\varphi$ dynamic, 111
  dynamic, 89, 111
  impedance, 89
  inductive, 89, 111
  three phase ($3\varphi$) Y-connected, 111
  linear, 111
  non-linear, 111
  power, 111
  resistive, 89, 111
  three phase ($3\varphi$) Y-connected, 111

magnetic fields, 15
Mallat theory, 22
mechanical resonances, 15
memory size, 100
metal-oxide-semiconductor field–effect
    transistor (MOSFET), 1
micro-processor technology, 12, 16

nodes, 40
  AC, 40
  capacitive, 40
  DC, 40
  inductive, 40
  input, 40
  output, 40
Nyquist rate, 28

operation
  direct sum, 19, 64
  inner product, 66
  operator, 20
  differential, 24
  fractional differential, 24
  inner product, 37
  linear, 20
  sparse, 20
  sampling, 29

Parseval's identity, 30
power, 1
  conditioner, 11
  converters, 1
  electronics, 1, 18
  flow, 40
  HVDC transmission, 5
  losses, 15
  quality, 1
  standards, 1
  supplies, 1
  systems, 1, 18

quarter-cycle symmetry, 15

resistance, 1
Riesz Bounds, 22, 63, 64
root-mean-square (RMS), 3

sampling, 18
  $N$-dimensional approximation, 37
  aliasing, 34, 35
  average period, 35
  constant rate, 33
  constant spacing, 33
  equidistant, 33
  interval, 28
  MRA support, 36
  non-uniform, 18, 34
    Beurling density, 35
    finite number of subsets, 34
    groups, 35
    recurrent, 18, 35
    relative separation, 34
  non-uniform recurrent
    MRA support, 66
    non-dyadic MRA support, 69
    sample's groups, 69

Nyquist rate, 34
perfect reconstruction, 34
spacings, 29
theorem, 18
uniform, 18, 29, 33
sampling theorem, 28
Kramer's extension, 29
mathematical approaches, 33
multi-dimensional approximation, 31
multiresolution analysis (MRA), 31
orthogonal projection, 29
perfect reconstruction, 28
scale, 66
initial value, 68, 110
time localization, 104
scale-time interval factor $\gamma$, 102, 104, 106
semiconductor switching elements, 1
sequence, 27
autocorrelation, 27, 67, 68
discrete, 28
dual filter, 27
non-equidistant samples, 34
non-zero, 28
signal, 18
$N$-dimensional approximation, 29
$N$-dimensional representation, 29
amplitude, 14
carrier
falling portion, 44
rising portion, 44
continuous-time (CT), 18
band-limited, 28
low pass, 28
MRA decomposition, 18
reconstruction, 45
series expansion, 74
discrete-time (DT), 32, 33
energy, 26
expansion, 20, 69
basis functions, 37
coefficients, 26
frequency components
time localization, 20
MRA expansion, 62
perfect reconstruction, 32
polynomial approximation, 34
processing, 18
conditions, 18
frequency-times analysis, 18

reconstruction, 18
sampling, 18
projection, 29
reference-modulating, 14
sampled form, 33
series expansion, 26
sinusoidal reference-modulating, 42, 106
discrete, 43
spline approximation, 34
simulation, 90, 111
solid-state technology, 12
space, 18
$L2$ square-integrable functions, 19
collection, 24
completeness, 76
continuous and bound functions, 29
dyadic structure, 28
high order, 23
Hilbert, 22
orthogonal basis, 22
linear, 19, 21, 23, 24
orthogonal, 24
linear span, 37
MRA nested, 19, 76
MRA scales, 37
MRA spanned, 19, 24
non-dyadic structure, *see* wavelet
non-orthogonal, 28
orthogonal complement, 19, 24, 37
orthogonal scaling, 19, 24
orthogonality, 26, 70
scaling, 37, 62, 70
spanning, 18, 19
wavelet, 37, 62, 70
subspace, 20
$N$-dimensional, 29
functions, 29
mapping, 29
orthogonal projection, 29–31
sampling, 30
orthogonal complement, 21
switching actions
time instants, 42
switching functions, 2
switching pulses, 13
duration (width), 13
location, 13

switching pulses (*cont'd*)
  pulse amplifier, 95, 119
  sampling frequency, 42
  sets of synthesis scale-based linearly-
    combined scaling functions, 108
  trains, 41
  volt-second average, 41
system of coordinates, 20

Tektronix TDS 3014B 4-channel digital
  phosphor oscilloscope, 95, 119
theory
  circuit, 1
  unified converter, 40
  wavelet, 65
train-of-impulses, 33
transforms, 18
  Z, 18
  continuous time Fourier (CTFT), 28
  discrete time Fourier (DTFT), 63
  Fourier, 18, 23
  generalized integral, 29
  Laplace, 18
transient operating conditions, 1
Turbo-*C* code, 95, 119

uninterruptible power supply (UPS), 1

voltage, 1
  $3\varphi$ line-to-line, 111
  $3\varphi$ line-to-neutral, 111
  hall-effect sensor, 95
  inverter output, 97
  load, 95
  THD$V$, *see* inverter, 90, 112

wavelet, 18
  basis functions, 18, 24
    design methods, 66
    MRA construction, 18
  bi-orthogonal, 26
  dyadic MRA, 38
  functions, 27
    applications, 62
    features, 62
  multiresolution analysis
  non-dyadic MRA, 65, 72
  orthogonal, 22
  orthogonal spaces, 24
  semi-orthogonal, 24
  shift-orthogonal, 27

# Books in the IEEE Press Series on Power Engineering

*Principles of Electric Machines with Power Electronic Applications,* Second Edition
M. E. El-Hawary

*Pulse Width Modulation for Power Converters: Principles and Practice*
D. Grahame Holmes and Thomas Lipo

*Analysis of Electric Machinery and Drive Systems,* Second Edition
Paul C. Krause, Oleg Wasynczuk, and Scott D. Sudhoff

*Risk Assessment for Power Systems: Models, Methods, and Applications*
Wenyuan Li

*Optimization Principles: Practical Applications to the Operations of Markets of the Electric Power Industry*
Narayan S. Rau

*Electric Economics: Regulation and Deregulation*
Geoffrey Rothwell and Tomas Gomez

*Electric Power Systems: Analysis and Control*
Fabio Saccomanno

*Electrical Insulation for Rotating Machines: Design, Evaluation, Aging, Testing, and Repair*
Greg Stone, Edward A. Boulter, Ian Culbert, and Hussein Dhirani

*Signal Processing of Power Quality Disturbances*
Math H. J. Bollen and Irene Y. H. Gu

*Instantaneous Power Theory and Applications to Power Conditioning*
Hirofumi Akagi, Edson H. Watanabe, and Mauricio Aredes

*Maintaining Mission Critical Systems in a 24/7 Environment*
Peter M. Curtis

*Elements of Tidal-Electric Engineering*
Robert H. Clark

*Handbook of Large Turbo-Generator Operation Maintenance,* Second Edition
Geoff Klempner and Isidor Kerszenbaum

*Introduction to Electrical Power Systems*
Mohamed E. El-Hawary

*Modeling and Control of Fuel Cells: Disturbed Generation Applications*
M. Hashem Nehrir and Caisheng Wang

*Power Distribution System Reliability: Practical Methods and Applications*
Ali A. Chowdhury and Don O. Koval

*Economic Market Design and Planning for Electric Power Systems*
James Momoh and Lamine Mili

*Operation and Control of Electric Energy Processing Systems*
James Momoh and Lamine Mili

*Restructured Electric Power Systems: Analysis of Electricity Markets with Equilibrium Models*
Xiao-Ping Zhang

*An Introduction to Wavelet Modulated Inverters*
S.A. Saleh and M. Azizur Rahman